A FORMULARY OF
PAINTS AND OTHER COATINGS
VOLUME I

Compiled by

Michael and Irene Ash

Chemical Publishing Co., Inc.
New York, N. Y.

©1978

Chemical Publishing Co., Inc.
ISBN 0-8206-0348-1 (pb)

Printed in the United States of America

PREFACE

The physical aspects of paint chemistry and technology are well covered in many up-to-date texts and the reader is strongly advised to become acquainted with this material before using this formulary.

This volume and the ones that will follow are a catalogue of formulas reflecting the state of the art in the paint industry. It is meant to serve as a tool for anyone who must turn out quality paint products, without the fiscal luxury of developing completely new, untested prototypes. However, this is not a cookbook. Most of the formulas contained are starting-point preparations that do require refining. It is that giant leap from nothing to a workable idea that this series attempts to make.

Anyone familiar with coatings knows it is impossible to have distinct chapters dealing with only one product. Of the chapters presented here almost all contain major elements of the others. The descending logical sequence of each chapter is loosely built on:

A. Use (primer, enamel, exterior paint, interior paint, luminescent paint, etc.) and special characteristics
B. Vehicle (oil, alkyd, epoxy, latex, etc.)
C. Color (if any specified)

Unless otherwise specified, all formulas have the quantities of ingredients given in parts by weight. Where test results were available, they follow the formula. A list of abbreviations that are used throughout the formulary is included.

All constituents appearing by their trademark name are printed in boldface type, and the manufacturers' names and addresses appear after the list of alphabetized tradenames in the Appendix.

ABBREVIATIONS

a	approximately
b.p.	boiling point
C	degrees Celsius
cm^2	centimeters squared
cps	centipoise(s)
CPVC	critical pigment volume concentration
F	degrees Fahrenheit
ft^2	square feet
g	gram(s)
gal	gallon(s)
hr	hour(s)
in.	inch(es)
in-lb	inch-pounds
KU	Krebs Units
lb	pound(s)
max	maximum
min	minute
Min	minimum
no.	number
NV	nonvolatile
%	percent
P/B	pigment to binder
PBW	parts by weight
pH	hydrogen ion concentration
PHG	per hundred gallons
PVaC	polyvinyl acetate
PVC	pigment volume concentration
rpm	revolutions per minute
s	second(s)
sol'n.	solution
theor.	theoretical
visc.	viscosity
VNV	vehicle nonvolatile
vol.	volume
Wt.	weight

CONTRIBUTORS

TABLE OF CONTENTS

Chapter I

PRIMERS

A primer is the first "gluelike", pigmented coat that acts as the interface between the finished coat and the substrate. It must be wettable and penetrable to achieve substrate adhesion, while ensuring compatibility with the finish coats.

Red Iron Oxide Silicone Primer

(Alkyd)

	lb	gal
Red Iron Oxide	259	6.90
Zinc Chromate	78	2.72
Talc	181	7.83
Varkyd 385-50X	503	61.01
Xylene	163	22.36
Bentone 38	2.5	.35
Exkin #2	.8	.11
Cobalt Naphthenate (6%)	2.5	.31
Lead Naphthenate (24%)	4.1	.42
PVC (%)		40
Nonvolatile (% Wt.)		45
Total NV by Wt.		64.7
Wt. gal		11.7
Visc. (KU)		70–80
Reduction: 15% with xylene		
Reduction Visc.:	#4 Ford cup	20–25

Basic Silico-Chromate Micaceous Iron Oxide Primer for Shop-Coat Application

(Alkyd)

		lb
1.	M-50 Basic Lead Silico Chromate	640
2.	#316 Micaceous Iron Oxide	275
	#38 Bentone Mastergel (15%)	37
3.	#325 Asbestine	73
	P-296-60 Soya Alkyd	225
	Raw Linseed Oil	248
	Mineral Spirits	33
	Troykyd — Cobalt Naphthenate (6%)	2.4
	Troykyd — Calcium Naphthenate (5%)	2.9
	Troykyd — Lead Naphthenate (24%)	6.0
	Solids by vol. (%)	78
	PVC (%)	36.4
	Wt. gal	15.4
	Visc. (KU)	84
	Nonvolatile (%)	90

Procedure:

Ball mill grind 1, 2, and 3 in vehicle then disperse "316" Micaceous Iron Oxide (no additional grinding).

Primer for Use in Conjunction with Micaceous Iron Oxide Corrosion-Resistant Top Coat

(Alkyd)

	lb
1681 Red Oxide	270
H.O. Zinc Oxide	46
1428 Zinc Chromate	46
400 Magnesium Silicate	100
1331-50 Med. Soya Alkyd	267
Mineral Spirits	55.5

Procedure:

Steel ball mill grind.

Let down:	lb
1331-50 Med. Soya Alkyd	245
Mineral Spirits	52
Xylol	6.5
Troykyd – Cobalt Naphthenate (6%)	1.5
Troykyd – Lead (24%)	6.0
Troykyd – Antiskinning Agent	1.5

White Metal Primer

(Alkyd)

	lb/100 gal
Easily Dispersible Rutile Titanium Dioxide	175.0
Talc (325 Mesh **Asbestine**)	210.0
Thixcin R	3.0
Haynie 111-70 Alkyd	380.0
Mineral Spirits	240.0
Cobalt (6%)	3.5
Zirconium (6%)	6.5
Antiskinning Agent	1.0
lb/gal	10.17
Visc. (KU)	82-87
Grind	5

General Purpose Flash Dry Primer

(Epoxy)

	lb	gal
Red Lead (97%)	94.9	1.26
Red Iron Oxide (**Mapico 516**)	94.9	2.21
Asbestine 3X	81.1	3.41
Nuosperse 657	2.7	.34
Eponol 53-L-32	346.2	39.89
MPA-60 (Xylene)	76.6	10.52
Roller mill grind		
CKR5254 Sol'n.	60.4	7.25
Methyl **Cellosolve**	119.8	14.93
Cyclo Sol 53	66.6	9.16
Xylene	79.9	11.03

Wt./gal/lb	10.2
Total Nonvolatile (%)	42.3
Flash Point (Pensky-Martin Closed Cup)	85°F

Procedure:

The pigment, extender, **Nuosperse 657**, **Eponol 53-L-32**, and **MPA-60** are dispersed on a 3 roll mill. This dispersion is then let down with the phenolic resin solution and the remaining solvent.

Zinc-Rich Primer

FORMULA NO. 1

(Epoxy)

Zinc Dust	1332.4
Araldite 488 N-40	410.7
Xylene	86.6
Methyl Isobutyl Ketone	43.3
Cellosolve	43.3
Pigment Suspension Agent	13.3

Procedure:

Add solvents to the resin solution, then add the zinc dust pigment with high speed agitation.

Nonvolatile Content (% by Wt.)	77.7
Nonvolatile Content (% by Vol.)	38.1
PVC (%)	60.5
Zinc Dust/Epoxy Resin Ratio (solids)	90/10
initial	100
after 2 weeks at 49 C (120 F)/no gassing	121
after 2 months at 25 C (77 F)/no gassing	106

Performance:

Cure Schedule, 25 C (77 F)	7 days
Substrate	Steel
Film Thickness, mils.	3–4
Dry Time, min	
Set-to-Touch	2
Paper Free	6
Dry Hard	11
Dry Through	13

Pencil Hardness, 1 day	HB-F
Flexibility, cylindrical mandrel	Pass ½ in.
Water Resistance, 25 C. (77 F) 3 months	No effect
Sea Water Resistance, 25 C. (77 F) 3 months	No effect
Salt Spray Resistance, 1000 hr	No. 6 med. dense blisters

No. 2

(Epoxy)

	lb/100 gal
38-407 Epotuf	350.0
Cab-O-Sil	8.0
Bentone 34	8.0
Zinc Dust No. 22	1570.0
Cobalt Naphthenate (6%)	0.8
Xylol	233.0

Note: Incorporation of pigment must be uniform.

Uses:

Automotive and industrial weld-through primers.

No. 3

(Epoxy)

	lb	gal
Zinc Dust No. 22	1618.8	27.2
Syloid AL-1	16.4	1.6
Thixatrol ST	11.0	1.3
Eponol 53-L-32	362.7	41.8
Methyl **Cellosolve**	130.0	16.2
Cyclo Sol 53	86.6	11.9

Procedure:
    Disperse the **Thixatrol ST** in the **Eponol** 53-L-32 Methyl Cellosolve and **Cyclo Sol 53** on a Cowles Dissolver (maximum thixotropy is developed at about 230–140 F). Slowly add the zinc dust and **Syloid A1-1** with good agitation. Continue mixing until zinc is thoroughly dispersed.

No. 4

(Epoxy)

	lb/100 gal
Araldite 488 N-40	408.0
Zinc Dust 430	1323.6
M-P-A 60 (Xylene)	13.2
Xylene	86.0
Methyl Isobutyl Ketone	43.0
Cellosolve	43.0
PVC (%)	60.5
Zinc Dust (%)	90.0
N.V. (% by Wt.)	77.7
N.V. (% by vol.)	38.1

Procedure:

Add solvents to the 488 resin solution. Then add the zinc dust pigment with high speed agitation.

Performance:

Initial Visc. (KU)	100	
Stability	2 weeks	2 months
	@ 120 F	@ 77 F
Visc. (KU)	121	106
Gassing	None	None
Dry Time (min)		
Set-to-Touch		2
Paper Free		6
Dry Hard		11
Dry Through		13
Pencil Hardness, 1 day		HB-F
Flexibility — ½″ Mandrel		Passed
Tabor Abrasion*		
Mg. Lost/1000 Cycles		135
Fresh Water Immersion —		
3 months		
Primer		
Blistering**		None

Primer and Topcoat	
Blistering**	4 Medium/None
Sea Water Immersion – 3 months	
Primer	
Blistering	None
	—White Spots—
Primer and Topcoat	
Blistering**	4 Medium
	Dense/None
Salt Spray Resistance	
1000 Hr	
Primer	
Blistering	6 Medium
	Dense/None
Blistering**	4 Medium/
	6 Few

*CS-10 wheel with 500 g load
**Along score/entier panel

No. 5

(Epoxy)

	lb/100 gal
Araldite 488 N-40	408.0
Zinc Dust 430	1323.6
M-P-A 60 (Xylene)	13.2
Xylene	86.0
Methyl Isobutyl Ketone	43.0
Cellosolve	43.0

No. 6

(Epoxy – Ester)

	lb/100 gal
Zinc Oxide XX50	257.0
Cab-O-Sil	7.5
Bentone 38	7.5
Epotuf 38-406 (6406-60)	318.0

	lb/100 gal
Mineral Spirits	313.0
Cobalt Naphthenate (6%)	1.3
Lead Naphthenate (24%)	3.3
Calcium Naphthenate (4%)	2.0
Exkin No. 2	0.8
Diethylene Triamine	1.0
Tetraethyl Orthosilicate	16.5

Pebble mill — 24 hr

Zinc Dust No. 22	1025.0

Uses:

Trade sales and maintenance corrosion resistant metal primer.

No. 7

(Epoxy — Ester)

	lb	gal
Epoxy Ester, DCO (50% solids in xylene)	335	42.7
Thixatrol ST	5	0.59
High Flash Naphtha	83	11.5
Xylene	83	11.5
Cobalt Napthenate (6%)	1	.13
Zinc Dust	1980	33.8

Procedure:

Disperse **Thixatrol ST** in the vehicle with high speed disc impeller (Cowles) at approximately 130 F. Add solvent and drier and then the zinc dust. The zinc dust is added using good and thorough agitation at a temperature below 130 F.

Application:

This primer can be applied by spray or brush.

White Brushing Primer

(Epoxy — Ester)

	lb/100 gal
Titanox RA	250.0
Nytal 300	318.0

	lb/100 gal
Epotuf 38-401 (6401-50)	230.0
Hi, Flash Naphtha	230.0

Pebble mill to 6 N.S.

Epotuf 38-401 (6401-50)	134.0
Carbitol	8.0
Cobalt Naphthenate (6%)	1.3
Lead Naphthenate (24%)	3.2
A.S.A.	0.9

Uses:

High quality trade sales and maintenance primer for metal, masonry and other surfaces.

Brown Aerosol Primer

(Epoxy)

		lb	gal
A.	Rutile Titanium Dioxide	100.0	2.92
	Carbon Black	10.0	0.69
	Strontium Chromate	5.0	0.15
	Red Iron Oxide	100.0	2.33
	Magnesium Silicate	80.0	3.40
	Litharge	4.0	0.05
	M-P-A 60 Xylene	14.4	2.00
	Soya Lecithin	8.0	1.00
	Epi-Tex 183	320.0	40.40
	Toluene	80.0	11.65
B.	**Epi-Tex 183**	186.0	23.40
	Lead Naphthenate (24%)	5.7	0.58
	Cobalt Napthenate (6%)	2.3	0.28
	Exkin #2 (Nuodex)	1.3	0.15
	Toluene	84.3	11.65

Procedure:

Steel ball mill A to 8–9 Paint Club Scale reading. Mill should reach 125 F. Then add B. Grind mix 1 hr and drain mill. Reduction viscosity — 11–13 s on #4 Ford Cup.

Reduction:

572.5	57.2	Base
310.2	42.8	Toluene

Can Loading:
  % by Vol.:
    61.6 Reduced Paint
    38.4 Propellant "P"

  % by Wt.:
    59.0 Reduced Paint
    41.0 Propellant "P"

Zinc Chromate Primer

(Epoxy)

Epoxy Mill Base:

Epon 1009 (or equivalent)	125
Polyvinyl Butyral, B-76-1 (10% in butyl acetate)	12
Titanium Dioxide, R-KB-2	26
Talc (325 mesh)	11
Zinc Chromate, X-2082	163
Blanc-Fixe 782	59
Bentone 27 (10% in ethylglycol acetate)	123
Ethyl Glycol Acetate	200
Xylol	76

Mondur Sol'n.

Mondur CB-60	205
	1000

% Solids	52
NCO/OH	1.02
Pigment/Binder Ratio	1/1

To reduce to spray viscosity, add 300 pbw of solvent blend: ethyl glycol acetate, butyl acetate, methyl ethyl ketone, and toluene (4:1:4:1 pbw).

Substrate Preparation:
The substrate to be primed must be free from rust, grease or dirt, Sand- or shotblasting to Steel Structures Painting Council-SP-563, white metal is preferred. Wire brushing, either hand or power, must be done with care. Pretreatment of the metal substrate with a two-package wash primer (polyvinyl butyral/zinc tetraoxychromate/phosphoric acid) is always advantageous and is mandatory in water immersion exposure.

Application:
This primer may be brush or spray applied and is particularly suited for finishes with good chemical resistance. The number of coats required depends on the sand blast profile of the substrate (1.5–2 mil profile is generally covered by one coat).

Zinc Dust Primer

FORMULA NO. 1

(Epoxy – Polyamide)

		lb/100 gal
A	Cab-O-Sil	26.5
	Epotuf 38-501 (6501-75)	150.0
	Cellosolve	53.0
	Xylol	40.0
	Three roll mill, one pass	
	MIBK	43.0
	Cellosolve	15.0
	Xylol	15.0
	Beckamine 31-510	9.5
B	Polyamide Hardener 37-618	73.0
	MIBK	28.0
	Cellosolve	28.0
	Xylol	6.0
C	No. 22 Zinc Dust	2372.0

Uses:
Shop primer for steel, general anticorrosive primer.

No. 2

(Phenoxy)

M-P-A 60 (Toluene) provides good pigment suspension properties without significantly increasing the viscosity of the system.

PKDA-8500	30
Polyketone 251	15
Zinc Dust Pigment	405
M-P-A 60	6.5
Solvent (blend: MEK/Butanol)	190
Toluene in ratio of 40:20:40)	

Procedure:

Predissolve the phenoxy resin and the ketone-formaldehyde resin in the given solvent mixture to give 40% N.V. solution. Disperse **M-P-A 60** (Toluene) in the vehicle solution, using a high speed disc impeller (Cowles). Next sift in the zinc dust pigment and thoroughly disperse in the vehicle. Add remaining solvent.

Visc. (Ford Cup #4) — s	24
Sag Control	Excellent
Pigment settling (1 week)	No settling
(1 month)	No settling
Zinc in dry film (%)	88.8
Nonvolatile (%)	70.6

Application:

This primer can be applied by brush or spray.

The Use of a Quick-Set Primer

In cases where previously painted walls, ceilings and other types of intersurfaces are in need or repair (e.g. spackling, plastering, etc.) before they are repainted, it is beneficial to use a primer over the patched areas. This primer must, however, be a suitable undercoat for latex emulsion or alkyd flat top coats by quickly forming the proper seal and thus preventing the penetration of these top coats into the repaired areas. If the porous substrates are not adequately sealed the binder portion of the top coat paint may be absorbed into these areas to some extent causing a sheen and/or color difference of the dried paint film over the patched area as compared to the other portion of the painted surface. Further, if such substrates were not primed it would be necessary to provide as many top coat applications as needed to give a completely uniform sheen and/or color effect over the entire painted wall or ceiling. Thus a good "quick set" primer is indicated particularly over repairs where a single, top coat application would be sufficient to cover over a previously painted substrate. In addition, when such primers are used, the primed surface can be painted over in about 30–35 min. A "quick set" emulsion primer of this type has particular value where alkyd top coats are being used since alkyd primed surfaces require a longer time to dry before the final coat can be applied.

Where new unpainted substrates are to be coated with latex emulsion or alkyd flat paints it is also important that such substrates be undercoated. Priming is of particular importance in dry wall construction where there is often a varying degree of porosity between the wall board and its taped seams (that of the joint cement itself). Therefore it is advantageous to prime the entire surface, wallboard and seam, to prevent sheen and/or color variations which may be obtained if such substrates were top coated without priming. However, in some cases priming the joint cement area only might be sufficient to give the desired results after the top coats are applied.

Quick-Set Primer

(PVAc)

	lb	gal
TriPure R-900	100	2.9
Atomite	250	11.1
Daxad-30	4	0.5
Butyl Cellosolve Acetate	24	3.0
Ethylene Glycol	28	3.0
Colloid 606	2	0.3
Troysan PMA-80	0.3	–
Igepal CO-610	2	0.2
Natrosol 250 HG (3%)	100	12.0
Water	281	33.7
Resyn 1255 (55% N.V.)	303	33.3
Visc.		70.8
PVC (%)		45.0

No guarantee is given or implied that some patent might not be infringed.

Exterior Primer

(PVAc)

	lb	gal
TiPure R-900	100	3.0
Basic Silicate White Lead	50	1.5
Drinkalite	75	3.3
Daxad 30	5	0.5
Colloid 606	2	0.3
Ethylene Glycol	28	3.0
Troysan PMA-30	4	0.4
Igepal CO-610	2	0.2
Natrosol 250 H.R. (3%)	50	6.0
Water	173	20.8
Flexbond 840 (56 N.V.)	555	61.0

PVC (%)	20.0
Consistency (KU)	70–75
Paint N.V. by vol. (%)	39.1
Paint N.V. by Wt. (%)	51.4
Vehicle N.V. by Wt. (%)	41.3

Use:

For new or unpainted wood surfaces.

Thixotropic Zinc Chromate Primer for Two-Coat Maintenance Systems

(Chlorinated Rubber)

Solids and Stabilizers:

Parlon S-10	28.68
Aroclor 1254	11.26
Aroclor 5460	7.08
Zinc Yellow (Imperial Color No. X2127)	12.52
Micro-Mica C-3000	13.43
Asbestine 325	12.52
Rutile Titanium Dioxide	4.90
Zinc Oxide (AZO-ZZZ-22)	6.35
Thixatrol ST	2.36
ERL-2774	0.72

Epichlorohydrin	0.18
Solvents	
Toluene	100.0
Solids as prepared (% Wt.)	55.0
Solids as applied (% Wt.)	42.0
Solids as applied (% vol.)	23.0

Procedure:

Disperse the **Thixatrol ST** in one-half the solvent at a temperature of 115 ±5°F. Mix approximately 5 min on a Cowles Dissolver. Add **Parlon** and other vehicle solids. Maintain temperature at 115°F. Add dry pigment after **Parlon** and resins are dissolved. Continue high speed mixing until desired fineness of grind is obtained. Add remainder of solvent.

Zinc Rich Primer

(Chlorinated Rubber)

Parlon S-20	4.04
Aroclor 1254	2.16
Aroclor 5460	1.08
Zinc Dust (high efficiency)	80.47
Thixatrol ST	.25
Xylene	6.60
High-Flash Naphtha	4.20
Turpentine	1.20
Vol. % of Zinc Dust in Dry Film	70
% Solids as prepared	88

Dry time in minutes — air-sprayed films (2.0 mils dry)

Set-to-Touch[1]	6
Cotton Free[2]	6
Foil-Free[3]	6
Through Dry[4]	15

Dry time in minutes — cast films (1.7 mils dry)

Set-to-Touch[1]	8
Cotton-Free[2]	8

Foil-Free[3]	8
Through Dry[4]	15

[1] Procedure described in Federal Test Methods Standard No. 141a. Method 4061.1, Section 4.2.2.

[2] Procedure from Federal Test Methods Standard No. 141a, Method 4061.1, Section 4.2.3.1 was applied to both sprayed and cast films.

[3] Determined by placing a 4-inch square of aluminum foil on the applied coating then rubbing the foil lightly with index finger and removing the foil. Any evidence of tack indicated incomplete drying.

[4] Determined by scratching the coating with a knife and observing the point at which the coating became hard and somewhat brittle in contrast to being plastic. This point is rather distinct when the PVC is as high as in zinc rich coatings.

General-Purpose or Elevated-Temperature Metal Primer

(Polyurethane)

Desmophen Mill Base:

Desmophen 1100	167
EAB 381-2 (10% in ethylglycol acetate)	17
TiO$_2$ (R-KB-2)	63
Talc (325 mesh)	17
Zinc Chromate X-2082	270
Blanc-Fixe 782	66
Bentone 27 (10% in ethylglycol acetate)	110
Ethyl Glycol Acetate	14
Xylol	30
Solvesso 100	30
Pentoxone	30

Mondur Sol'n:

Mondur CB-60	186
	1000
% Solids	70
NCO/OH	0.72
Pigment/Binder	1.5/1

To reduce to spray viscosity, add 1 part of solvent mixture (ethyl acetate, butyl acetate, ethyl glycol acetate, toluene 1:1:1:1 pbw) to 2 parts of mixed primer.

For exposures at normal temperatures, the zinc chromate noted in the formulation is used. For long-term exposure to temperatures up to 200 C, strontium chromate should be used as the anticorrosive pigment. Where temperatures higher than 100 C are encountered, red iron oxide is adequate as the prime pigment.

Substrate Preparation:

The substrate must be free from rust, grease and dust. Sandblasting or gritblasting is the preferred method of surface preparation. Rust removal by hand or power brush should be carried out with great care. Pretreatment of the substrate with a two-package wash primer (polyvinyl butyral/zinc tetraoxychromate/phosphoric acid) is always an advantage.

One-Component Metal Primer

(Polyurethane)

Multron R-12A	152.0
Butvar B-76-1	26.0
Talc	53.0
Zinc Chromate X-2082	13.0
TiO_2 (R-KB-2)	95.5
Bentone 27	2.5
Ethyl Glycol Acetate	164.0
Methyl Ethyl Ketone	244.0
Toluene	82.0
Mondur S	168.0
Solids (%)	45
NCO/OH	1.1
Pigment/Binder Ratio	0.7/1

Curing Schedules:

60–90 s	@ 500°F
5 min	@ 400°F
15 min	@ 350°F
30 min	@ 300°F

This primer system is characterized by high resistance to detergents and salt spray. It is recommended for evaluation as a primer in coil coating and appliance applications.

Universal Metal Primers

FORMULA No. 1

(Methyl Vinyl Ether)

Gantrez VC	19.87
Oncor M-50	36.64
Bentone 38	0.16
Toluene	36.95
VM&P Naphtha	3.19
Solvesso 100	3.19
PVC	35.22
Theoretical 1-mil coverage	539.71 ft^2/gal
Nonvolatiles by Wt. (%)	56.86
Nonvolatiles by vol. (%)	33.65
Visc. (KU)	92
Wt. per gal (lb)	10.97
Specific gravity	1.32

No. 2

(Methyl Vinyl Ether)

Gantrez VC	9.51
Chlorinated Paraffin (40%)	1.90
Red Lead	64.90
Bentone 38	0.04
Toluene	17.65
VM&P Naphtha	3.00
Solvesso 100	3.00

Chlorinated paraffin used as plasticizer; high PVC value

PVC (%)	43.64
Theoretical 1-mil coverage	608.16 ft^2/gal
Nonvolatiles by Wt. (%)	74.45
Nonvolatiles by vol. (%)	37.92
Visc. (KU)	94
Wt. per gal (lb)	18.69
Specific gravity	2.24

Exterior Primer for Nonstaining Woods

(Vinyl-Acrylic)

Water	150
Tamol 731	7
Triton X102	2
NXZ	1
Ethylene Glycol	25
Pine Oil	3
PMA-60	2
Natrosol 250 MR	2
Ti-Pure R-610	150
Camel Carb	155
W. G. Mica	35
ParCryl 300	650
NXZ	1
Ammonium Hydroxide (28%)	2
Yield	100 gal
PVC (%)	30
Solids	58.1
Visc. (KU)	76–82
pH	9.2–9.8

Chapter II

EXTERIOR PAINTS

Paints with a low pigment load usually have exterior durability. The large binder content provides the film matrix required to withstand exterior conditions. The pigment volume concentration (PVC) is usually well under 40 percent.

The low PVC (i.e. high binder) results in a strongly bound, highly resistant film possessing the following variable qualities: good weathering, glossiness, resistance to specific chemicals and abrasion. The formulas presented in this chapter possess one or more of these characteristics to a greater or lesser degree.

Rapid-Drying Stain and Blister-Resistant House Paint

(Alkyd)

	lb	gal
Rutile Titanium Dioxide	40	1.14
Anatase Titanium Dioxide	160	4.92
Micronized Magnesium Silicate	200	8.44
Thixcin R	4	0.12
Phenyl Mercury Propionate	3	—
Calcium Naphthenate (4%)	11.5	1.5
Zinc Naphthenate	4.0	0.5
Aroplaz 1254-M-70	382	48.00
Aroplaz 1254-M-70	49	6.00
Heavy Mineral Spirits	196	29.12
Cobalt Naphthenate (6%)	1	0.13

Antiskinning Agent	1	0.13
PVC (%)		30
Vehicle Nonvolatile (%)		48.3
Gloss		40–50
Visc. (KU)		75–85

Exterior White Paint

(Alkyd)

	lb	gal
52R13 Resin (60% NV)	164.0	20.00
Amsco 460	80.0	12.00
Antiskinning Agent	1.0	0.13
Rutile Titanium Dioxide	250.0	7.17
Anatase Titanium Dioxide	50.0	1.54
Uni-Mal 303*	225.0	10.50
Chlorowax 70	100.0	7.25
Thixcin R	3.0	–

Disperse with high speed agitation to #5 grind. Then add:

White Refined Linseed Oil	78.0	10.00
52R13 Resin (60% NV)	236.0	31.50
Cobalt Drier (6%)	2.0	0.25
Zirconium Drier (6%)	5.0	0.75
Phenyl Mercuric Oleate (30%)	15.00	2.00

lb/gal	11.7
PVC (%)	30
Vehicle Nonvolatile (%)	61
Visc. (KU)	85–90

Flat Exterior Paint

(Alkyd)

	lb	gal
Rutile Titanium Dioxide	200	5.72
Acicular Zinc Oxide	150	3.25
Magnesium Silicate	150	6.33
Wet Ground Mica	50	2.11

	lb	gal
Puffing and Anti-Sag Additive	8	0.96
Aroflat 3025-P-40	435	60.90
Heavy Mineral Spirits	138	20.65
Lead Naphthenate (24%)	3	0.31
Managanese Naphthenate (6%)	1	0.12
Antiskinning Agent	1	0.13
PVC (%)		46.0
Vehicle Nonvolitale (%)		30.4
lb/gal		11.30
		90–100

Has very good application and nonpenetration properties with good tint retention and mildew resistance. However, like all exterior flats, durability is not as good as conventional linseed oil paints. Where a flat exterior is desired we strongly recommend using this over a suitable primer for both new and repaint work for good durability.

In severe mildew areas use of phenyl mercury preservative is suggested.

Exterior Alkyd Paint

	FORMULA NO. 1 (Primer)		NO. 2 (Top Coat)	
	lb	gal	lb	gal
Ti-Pure R-900	100	2.9	200	5.8
Drikalite	350	15.6	375	16.7
Water Ground Mica	50	2.1	50	2.1
Thixcin	4	0.5	4	0.5
Metasol 57	4	0.2	4	0.2
Aroflat 3025 (40 NV)	450	63.0	372	52.0
Cobalt Naphthenate (6%)	1	0.1	1	0.1
Zirconium Naphthenate (6%)	4	0.5	4	0.5
Antiskinning Agent	1.5	0.2	1.5	0.2
Mineral Spirits	97	14.9	143	21.9
PVC (%)		49		58
Visc. (KU)		85		85

No. 3

	lb	gal
Disperse:		
Varkyd 505-60	233	29.9
Mineral Spirits	67	10.2
Calcium Naphthenate (4%)	3	.4
Titanium Dioxide	150	4.3
Calcium Carbonate — **Snowflake**	400	20.4
Celite 281	90	4.6
Aluminum Stearate	7.5	1.0
Let Down:		
Varkyd 505-60	140	17.9
Mineral Spirits	80	12.1
Cobalt Naphthenate (6%)	2.25	.3
Lead Naphthenate (24%)	4.5	.5
Antiskinning Agent — **Exkin #2**	.5	.1
Yield (gal)		100
Wt./gal (lb)		11.5
PVC (%)		51
Vehicle Nonvolatile (Wt.)		43
Total NV (% Wt.)		74
Visco. (KU)		80

No. 4

	lb
Rutile Titanium Dioxide	200
Magnesium Silicate	250
Wet Ground Mica	65
Aroflat 3025-P-40	435
Heavy Mineral Spirits	121
Phenyl Mercury Preservative	8.7
Lead Naphthenate (24%)	3
Manganese Naphthenate (6%)	1
Antiskinning Agent	1

Green Trim Paint

(Alkyd)

Medium Chrome Green	6.05
Talc, Medium Fine	2.00
Bentone 34[1]	0.20
Methanol	0.14
Q-Bodied Linseed Oil	12.93
Alkyd Resin (**Beckosol P-296-70**)[2]	14.12
Alkyd Resin (**Beckosol P-297-70**)	39.71
Mineral Spirits	21.25
Lead Naphthenate (24%)	1.23
Cobalt Naphthenate (6%)	1.13
Antiskinning Agent	0.14
Phenyl Mercury Oleate (10% Hg.)	1.10

[1] thickener and flow control. Methanol is used to activate it.
[2] 70% nonvolatile. Wt.: 9.61 lb per gal PVC: 14.5%. Binder: 75% alkyd resin, 25% bodied linseed oil.

Outside White House Paint

FORMULA NO. 1

(Alkyd)

	lb/100 gal
Anatase Titanium Dioxide	137
Rutile Nonchalking Titanium Dioxide	33
Asbestine 3X Talc	255
Thixcin R	2
Haynie 121-50 Alkyd	318
Haynie Admiral X-Y	160
Mineral Spirits	126
Cobalt (6%)	3.3
Calcium (5%)	3.3
Lead (24%)	6.5
Antiskinning Agent	1
PMO-30 Mildewcide	10
lb/gal	10.55
Visc. (KU)	87–95

	lb/100 gal
Grind	3
PVC (%)	30

No. 2

	lb/100 gal
Rutile Titanium Dioxide	4.28
Anatase Titanium Dioxide	1.54
Litharge	0.05
Barium Metaborate	3.53
Talc (Fine-Textured)	2.11
Talc (Super-Fine)	4.24
Bentone 34, Thickener	0.47
Alkyd Resin Sol'n 3490	48.38
Alkyd Resin Sol'n 2964	9.00
Mineral Spirits	24.07
Manganese Naphthenate (6%)	0.16
Zirconium Drier (6%)	1.31
Methyl Alcohol	0.36
Thickening Agent **RL-80**	0.50

Hi-Hiding Gloss White House Paint

(Alkyd)

	lb	gal
Titanox RA-50	300	8.6
Atomite	250	11.1
Thixcin R	3	0.4
Slow Mineral Spirits **Std. 410**	47	7.0
Varkydol 210-100	203	25.0
Varkydol 210-100	244	30.0
Standard 410	136	20.0
Cobalt Naphthenate (6%)	3	0.4
Calcium Naphthenate (4%)	8	0.9
Volatile ASA	1	0.1
PMO-10 Mildewcide	9	1.1

PVC (%)	26.2
Vehicle Nonvolatile (% Wt.)	71.0
Total Nonvolatile (% Wt.)	83.0
Wt./gal (lb)	11.48
Visc. (KU)	75–80
Brushing	Good–Very Good
Leveling	Fair
Sagging Resistance	Excellent
Settling	None
Gloss	Excellent
Dry — Set-to-Touch	3–5 hr
Dry Hard	Overnight

White Primer

(Linseed Oil)

Basic Carbonate White Lead	26.0
Basic Sulfate White Lead	24.5
Rutile Titanium Dioxide — Nonchalking	12.3
Talc	37.2
Raw Linseed Oil	24.0
Bodied Linseed Oil	45.0
Mineral Spirits	29.4
Lead Naphthenate (14%)	1.1
Manganese Naphthenate (6%)	0.5

Exterior White House Paint

(Linseed Oil)

Anatase Titanium Dioxide	100
Rutile Titanium Dioxide	100
Leaded Zinc Oxide	150
Calcium Carbonate	600
Bodied Linseed Oil (Z_3)	135
Refined Linseed Oil	265
Mineral Spirits	125
Lead Naphthenate (24%)	8
Manganese Naphthenate (6%)	1
Antiskinning Agent	1

Stain and Blister Resistant House Paint

(Long Oil)

	lb	gal
Free Chalking Titanium Dioxide	100	3.09
Nonchalking Titanium Dioxide	100	2.86
Magnesium Silicate	300	12.65
Thixcin R	5	0.60
Admerol 75	337	41.30
Petroleum Solvent	221	33.00
Fungicide	17	2.10
Manganese Naphthenate (6%)	2	0.25
Lead Naphthenate (24%)	5	0.50
Antiskinning Agent	1	0.13

Procedure:

Grind. One pass through loose roller mill. Care should be taken not to mill at temperatures exceeding 130 F to prevent seeding of **Thixcin**.

PVC (%)	31.0
VNV (%)	60.5
Av. lb per gal.	11.30
Visc. (KU) Fresh	75
1 week	88
2 weeks	91

Driers: 0.036% Manganese
0.35% Lead as metal on vehicle solids

Calcium Carbonate Pigment in Water-Soluble Linseed Oil Paints

The introduction of a water-soluble linseed oil paint vehicle has stimulated a great deal of interest in the paint industry. This interest was created by the novel approach of having the proven durability of linseed oil combined with the desired convenience of water cleanup.

One of the advantages of this type of water-soluble linseed oil system is that it can be formulated at higher solids volume than most resin emulsion type systems; however, not as high solids as is true for conventional linseed oil paints. Even higher solids can be obtained in Linaqua systems by the incorporation of ground calcite extender pigments. Thus the use of such extenders should give longer film life.

A lower amount of zinc oxide is required in calcium carbonate containing paints to adequately control mildew growth than is necessary when such products are formulated with other commonly used grads of extender pigments. Therefore, this is an additional economic advantage when ground calcite extenders are used.

A few outstanding features which Linaqua type paints display are 1) ease of application (i.e. as compared to conventional linseed oil paints), 2) high gloss, and 3) easy cleanup of brushes and tools with water. These are properties which are quite appealing to the "do it yourself" consumer.

Water-Soluble Linseed Oil Paints

	FORMULA NO. 1 (Self-Cleaning – White)		NO. 2 (Chalk-Resistant – White)		NO. 3 (Pastel – Tint Base)	
	lb	gal	lb	gal	lb	gal
TiPure R-900	50	1.4	200	5.8	100	2.9
TiPure FF	150	4.6	–	–	–	–
XX631 (ZnO)	100	2.1	250	5.4	100	2.1
Drikalite	325	13.5	250	11.1	400	17.8
Linaqua (85% N.V.)	332	41.0	324	40.0	324	40.0
Lead Naphthenate (24%)	7	0.7	7	0.7	7	0.7
Manganese Naphthenate (6%)	0.9	0.1	0.9	0.1	0.9	0.1
Cobalt Naphthenate (6%)	2.8	0.4	2.8	0.4	2.8	0.4
Water	294	35.2	304	36.5	300	36.0
PVC (%)		39.8		40.0		40.6
Visc. (KU)		85		85		85

	No. 4 (One-Coat White House Paint)		No. 5 (Chalk-Resistant White House Paint)	
	lb	gal	lb	gal
TiPure R-900	300	8.8	200	5.8
XX 631 Zinc Oxide	250	5.3	250	5.3
Drikalite	400	17.8	500	22.2
Refined Linseed Oil	247	32.0	247	32.0
Z_3 Bodied Linseed Oil	128	16.0	128	16.0
Aluminum Stearate-GM	2	0.2	2	0.2
Lead Naphthenate (24%)	8	0.8	8	0.8
Manganese Naphthenate (6%)	1	0.1	1	0.1
Antiskinning Agent	1	0.1	1	0.1
Mineral Spirits	123	18.9	107	16.5
PVC (%)	40.5		40.5	
Visc. (KU)	85		85	

No. 6

(Self-Cleaning White House Paint)

	lb	gal
TiPure R-900	50	1.6
TiPure-FF	150	4.6
XX631 Zinc Oxide	100	2.1
Duramite	600	26.7
Drikalite	—	—
Refined Linseed Oil	247	32.0
Z_3 Bodied Linseed Oil	128	16.0
Aluminum Stearate-GM	2	0.2
Lead Naphthenate (24%)	8	0.8
Manganese Naphthenate (6%)	1	0.1
Antiskinning Agent	1	0.1
Mineral Spirits	102	15.8
PVC (%)	42.3	
Visc. (KU)	85	

Alkyd-Modified Acrylic House Paint with Zinc Oxide

FORMULA NO. 1

	lb	gal
Premix:		
Water	200.0	24.00
AMP-95	4.0	0.51
Potassium Tripolyphosphate	1.0	—
Defoamer	2.0	0.26
Disperse pigments:		
AZO-33 Zinc Oxide	100.0	2.14
Titanox 2062	200.0	6.08
Titanox 1080	50.0	1.54
Nytal 300 Talc	100.0	4.31
Grind; then add:		
Preservative ⎫ Premix	0.5	0.05
Ethylene Glycol ⎭	50.0	5.40
Rhoplex AC-388 Acrylic		
Emulsion (50%)	314.0	35.70
Aroplaz 1271 PL Alkyd Resin (100%)		
with Driers	35.0	4.20
Tributyl Phosphate	8.0	0.95
Defoamer	2.0	0.26
Water	119.8	14.38
Triton CF-10 Surfactant	2.0	0.22

Visc. (KU) at 77 F — Fresh	71
Overnight	73
1 week	75
1 month	78
1 month at 125 F	88
pH — Fresh	9.6
1 month at 125 F	9.4
PVC (%)	40.2
Total Solids (% by Wt.)	54

No. 2

	lb	gal
Premix:		
Water	100.0	12.00
AMP-95	4.0	0.51
Triton X-100 Surfactant	2.0	0.25
Increase speed and add slowly:		
Aroplaz 1271 PL (100%) with Driers	23.4	2.80
Disperse pigments:		
E-P AAA 427W Zinc Oxide	100.0	2.14
Titanox 2062	225.0	6.84
Duramite	65.0	2.88
Grind; then add:		
Ethylene Glycol ⎫ Premix	18.5	2.00
Preservative ⎭	0.5	–
Cellosize QP-15000 (2%)	208.3	25.00
Defoamer	2.0	0.26
Rhoplex AC-388 (50%)	280.0	31.84
Water	112.3	13.48

Visc. (KU) at 77 F – Fresh	77	
Overnight	80	
2 weeks	99	
13 months	99	
pH – Fresh	9.4	
PVC (%)	40.2	
Total Solids (% by Wt.)	48.5	

No. 3

	lb	gal
Premix:		
Water	150.0	18.00
AMP-95	5.0	0.64
Potassium Tripolyphosphate	1.0	–
Defoamer	2.0	0.26

Increase speed and add slowly:

Chemacoil TA-100 Resin (100%)	34.5	4.42

Disperse pigments:

AZO-33 Zinc Oxide	100.0	2.14
Ti-Pure R-966 Titanium Dioxide	200.0	6.08
Asbestine 3X Talc	150.0	6.51

Grind; then add slowly:

Rhoplex AC-388 Acrylic Emulsion (50%)	332.0	37.75
Defoamer	2.0	0.26
Tributyl Phosphate ⎫	8.0	0.99
Ethylene Glycol ⎬ Premix	50.0	5.38
Skane M-8 Preservative ⎪	2.0	0.23
Cobalt Drier (6%) ⎭	1.0	0.12
Cellosize QP-4400 (2½%) Thickener	141.7	17.00
Triton CF-10 Surfactant	2.0	0.22

Visc. (KU) at 77 F — Fresh		85
Overnight		86
1 week		87
1 month		91
1 month at 125 F		112
pH — Fresh		9.8
1 month at 125 F		9.4
PVC (%)		40.0
Total Solids (% by Wt.)		55.1

Exterior Ranch Red House Paint

(Alkyd – Acrylic)

Igepal CTA-639 effectively emulsifies alkyd and oil additives used to improve adhesion of exterior paints to chalky or smooth surfaces. It is unique in stabilization of alkyd-water emulsions to produce excellent clean up, can stability, and color development in this type of paint.

	lb	gal
Charge into mixer under agitation:		
Water	101.0	12.2

	lb	gal
Daxad 23	1.0	0.1
Tamol 731	4.0	0.4
Colloid 677	1.5	0.2
Igepal CTA-639	3.0	0.3
PMA-30	0.5	—
Natrosol 250 HR (1.5%)	50.0	5.9
R-1599 Pure Iron Oxide	125.0	3.0
Ethylene Glycol	30.0	3.2
Barytes No. 1	110.0	3.0
ASP 400	95.0	4.4
Celite 281	15.0	0.8
Carbitol Acetate	10.0	1.2

Disperse in high-speed mill, then add slowly with agitation:

	lb	gal
Natrosol 250 HR (1.5%)	230.0	27.5
Resyn 2243	245.0	26.9
Premix (then add):		
Igepal CTA-639	5.0	0.6
Aroplaz 1271	45.0	5.3
PMO-30	4.0	0.5
Advacar (6% Cobalt Drier)	0.5	—
Advacar (24% Lead Drier)	1.5	0.2
Colloid 677	1.0	0.1
Water	35.0	4.2

Note: Protect thickener solution with 0.08% PMA-30 if stored prior to use.

Solids (%)	48.7
PVC (%)	37.0
CPVC (%)	45.0
Wt./gal (lb)	11.1
Visc. (KU)	80–85
85° Angular Sheen	1.0
Contrast Ratio	1.0
% Alkyd Modification	25%

Flat Exterior White Paint

(Vinyl – Acrylic)

	lb	gal
Premix:		
Water	250.0	30.00
AMP-95	5.0	0.64
Tetrapotassium Pyrophosphate	2.0	—
Disperse pigments:		
Tronox CR-801 Titanium Dioxide	150.0	4.40
St. Joe 17 Zinc Oxide	100.0	2.14
Micro-White 50 Calcium Carbonate	150.0	6.64
Grind; then add:		
Ethylene Glycol	20.0	2.14
Cellosize QP-4400 Thickener ⎫ Premix	3.0	0.24
Tris Nitro (50%) Preservative ⎭	0.5	0.05
Amsco Res 3011 Vinyl Acetate/		
Acrylic Copolymer Emulsion (55%)	327.6	36.00
Defoamer	3.0	0.41
Igepal CO-630 Surfactant	3.0	0.34
Water	141.6	17.00

Visc. (KU) at 77 F – Fresh	72	
Overnight	73	
1 month	73	
1 month at 125 F	85	
pH – Fresh	6.5	
1 month at 125 F	6.5	
PVC (%)	40.9	
Total Solids (% by Wt.)	50.1	

Exterior Flat

FORMULA No. 1

(Acrylic)

	lb/100 gal
Water	125
Tamol 731 (25%)	8

	lb/100 gal
Polyglycol P-1200	8
Dowicil 100	1
Ti-Pure R-901	200
Atomite	257
Ethylene Glycol	15
Triton X-100	5
Rhoplex AC-35 (46%)	387
Water	170
XD-7845.01	4

No. 2

(Acrylic)

	lb/100 gal
Water	125
Tamol 731 (25%)	8
Polyglycol P-1200	8
Dowicil 100	1
Ti-Pure R-901	200
Atomite	257
Ethylene Glycol	15
Rhoplex AC-34 (46%)	387
Water	170
XD-7845.01	4

Exterior White Paint

FORMULA No. 1

(Polyvinyl Acrylic)

	lb	gal
Pigment Grind:		
Water	298.0	35.8
Tamol 731	8.0	0.9
Potassium Tripolyphosphate	0.3	—
Igepal CTA-639	3.0	0.3
Dowicil S-13	5.0	0.5

	lb	gal
Ethylene Glycol	20.0	2.2
Ucar Filmer 351	10.0	1.3
Defoamer[1]	5.0	0.7
Cellosize Hydroxyethyl **Cellulose**		
QP-4400H	5.0	0.5
Rutile Titanium Dioxide[2]	175.0	5.3
Anatase Titanium Dioxide[3]	50.0	1.6
Mica 325 mesh W.G.	25.0	1.1
Talc[4]	125.0	5.3

Let down:

	lb	gal
Ucar Latex 365	400.0	44.5

PVC (%)	36.4
Total Solids	54.0
Visc. (KU)	85–90
Wt. per gal (lb)	11.3

[1] Bubble Breaker 746; Colloid 585
[2] Ti-Pure R-966; Titanox RA-46; Unitane OR-650
[3] Ti-Pure LW; Titanox 1000; House Head A-420
[4] Nytal 300; Asbestine 325

Exterior White Latex House Paint

(Vinyl Acrylic)

	lb	gal
Charge into tank and mix until uniform:

	lb	gal
Water	293.0	35.2
Nopcosant K	8.0	0.8
KTPP	0.3	—
Tenlo 100	2.0	0.2
Nopcocide N-96	6.0	0.4
Ehtylene Glycol	20.0	2.2
BEP (1-Butoxyethoxy-		
2-Propanol)	15.0	1.7
2-Ethylhexyl Acetate	5.0	0.7
Foamaster B	2.0	0.2

Add the following and grind with a Cowles type mixer for 20 min (3800–4500 fpm):

Cellosize QP 4400	5.0	0.5
Ti-Pure R-900	175.0	5.0
Ti-Pure FF	50.0	1.5
Mica (water ground 325 mesh)	25.0	1.1
Asbestine 325	125.0	5.3

Reduce speed to 1900 fpm and add:

Ucar Latex 360	400.0	44.5
Foamaster B	3.0	0.3

No. 2

(Acrylic)

	lb	gal
Premix:		
Water	125.0	15.00
Daxad 30 Dispersant	8.0	0.94
Tergitol NPX Surfactant	4.0	0.45
Victawet 35-B Surfactant	2.5	0.26

Increase speed and add slowly:

Chemacoil TA-100	74.0	9.48

Adjust speed to disperse pigments:

Azo ZZZ-33 Zinc Oxide	75.0	1.61
Titanox RA-NC Titanium Dioxide	175.0	5.00
Titanox A-168-LO Titanium Dioxide	25.0	0.77
Asbestine 3X Talc	100.0	4.21
Ethylene Glycol	18.5	2.00
Nuodex PMA-18 Fungicide	3.0	0.26
Nopco NDW Defoamer	4.0	0.53

Slow to mixing speed:

Cellosize QP-15000 (2%) Thickener	165.0	19.80
Rhoplex AC-34 Acrylic Emulsion (46%)	372.0	41.74
Super-Cobalt Drier	1.0	0.13

Visc. (KU) — Fresh	80
Overnight	80
8 months	88
PVC (%)	29.8

Chemacoil TA-100 = 30% of binder solids

No. 3

(Acrylic)

	lb	gal
Premix:		
Water	100.0	2.00
Daxad 30 Dispersing Agent	8.0	0.94
Tergitol NPX Surfactant	3.6	0.41
Victawet 35-B Surfactant	2.5	0.26
Cellosize QP-15000 (2%) Thickener	41.6	5.00
Increase speed and add slowly:		
Chemacoil TA-100	65.0	8.33
Adjust speed to disperse pigments:		
Titanox RA-NC Titanium Dioxide	200.0	5.72
Asbestine 3X Talc	150.0	6.32
Nopco NDW Defoamer	4.0	0.53
Ethylene Glycol	18.5	2.00
Nuodex PMA-18 Fungicide	8.0	0.72
Slow to mixing speed:		
Cellosize QP-15000 (2%) Thickener	166.6	20.00
Rhoplex AC-34 Acrylic		
Emulsion (46%)	330.0	37.06
Super-Cobalt Drier	1.0	0.13
Water	16.6	2.00

Visc. — Fresh	83 KU
Overnight	84 KU
12 months	84 KU
PVC (%)	35.3

Chemacoil TA-100 = 30% of binder solids

Exterior White Paint — Alkyd Modified

(Vinyl Acrylic)

	lb	gal
Pigment Grind:		
Water	250.0	30.0
Merbac 35	0.5	—
Ethylene Glycol	25.0	2.7
Ucar Filmer 351	10.0	1.3
Strodex MOK	5.5	0.6
Tamol 850	8.0	0.8
Igepal CTA-639	2.0	0.2
Colloid 677	1.0	0.1
Cellosize Hydroxyethyl Cellulose QP-4400H	4.5	0.4
Nopcocide N-96	5.0	0.4
Titanox 2020	225.0	6.6
Zinc Oxide	50.0	1.1
Snowflake	50.0	2.2
Celite 281	25.0	1.3
Nytal 300	50.0	2.1
Let Down:		
Ucar Latex 365	305.0	33.8
Igepal CTA-639	8.0	1.0
Varkyd 515-100 ⎫	41.5	5.0
Cobalt Drier (6% Co) ⎬ Premix	1.5	0.2
Zirconium Drier (6% Zr) ⎭	3.0	0.3
Colloid 677	1.5	0.2
Water	81.0	9.7

PVC (%)	38.0
Total Solids (%)	55.0
Angular Sheen (85°)	7
Brightness (%)	89
Visc. (KU)	80–85
Wt. per gal (lb)	11.5
Contrast Ratio	0.97
Alkyd Modification	20

White Exterior High Solids Acrylic Latex Paint

	lb/100 gal
Grind:	
Water	250
Pigment Dispersant (25%)	9.2
Dow Polyglycol P 1200	2.0
Rutile Titanium Dioxide	185
Anatase Titanium Dioxide	45
Mica	30
Calcium Carbonate	100
Clay	55
Dowicil* S-13 Antimicrobial	6

Grind for 15–20 min or until proper dispersion is achieved.

Ethylene Glycol ⎫	added as a slurry	⎧ 17.5
Methocel K15M† ⎭		⎩ 3.5
Defoamer		5.0

Mix for 10 min

Let down:

Acrylic Copolymer Latex (47%)	450

Mix for 10 min or until smooth:

PVC (%)	40
Nonvolatile	54.9
pH	Adjust to 8.5–9.0
Visc. (KU)	85–90

†Methocel **J20MS, J12MS, K15MS** may be substituted. If so, the pH must be above 8.5 to completely break the cross link needed for good dispersion.

Java Brown Exterior Finish

(Vinyl-Acrylic)

Water	174
Tamol 731	9
CF-10	2
PMA-60	2

NXZ	1
Ethylene Glycol	25
Natrosol 250 MR	3
Mapico 418	80
Mapico 420	20
Gold Bond R	300
Pine Oil	3
ParCryl 300	515
NXZ	1
Ammonium Hydroxide (28%)	2
Yield (gal)	100
PVC (%)	40.0
Solids	56.7
Visc. (KU)	87–93
pH	9.4–9.8

Mildew-Resistant Exterior White Paint

(Acrylic)

	lb
Titanium Dioxide **R-901**	175
Asbestine 325	130
Water	110
Zinc Oxide #17	75
Titanium Dioxide FF	25
Ethylene Glycol	20
Carbitol	8
Victawet 35B	6
Tamol 731	5
Igepal 630	5
Colloids 581-B	2
Potassium Tripolyphosphate	1
Rhoplex AC-34	375
Cellosize QP-15000 (3%)	120
Castung 235 Premix*	105
Water	20
PVC (%)	35
Solids (%)	54
Visc. (KU)	78

Castung 325 Premix	lb
Castung 235	550
Igepal 630	10
PMO-10	15
Advacar Cobalt (6%)	8
Advacar Lead (24%)	20
Water	397

Baker Comment:

The **Castung 235** modified latex paints were applied over oil primer and exposed after August 15, which is considered the most difficult time of the year. No intercoat peeling or blistering occurred even on the bottom edge of the panels after 3 years of exposure.

Exterior Deep Green — Modified with Alkyd

(Vinyl — Acrylic)

Water	115
Tamol 731	7
Triton CF-10	2
Nopco NDW	1
Metasol 57	1.8
Chromium Oxide	100
Imsil A-25	300

Mix on a high speed mill and add:

Texanol	7
Ethylene Glycol	35
Natrosol 250 MR	2.5
T&W 345-100 Super Alkyd	30
Lead (25%)	1
Manganese (6%)	0.5
ParCryl 400	405
Ammonium Hydroxide	2
Water	110
Yield (gal)	100
PVC (%)	39
Solids	57
Visc. (KU	74–80
pH	9.5

Semigloss Emulsion Paint

(Acrylic)

	lb	gal
Premix:		
Water	125.0	15.00
Triton X-100 Surfactant	6.0	0.68
Daxad 30 Dispersing Agent	5.0	0.52
Potassium Tripolyphosphate	2.0	0.10
Increase speed and add slowly:		
Chemacoil TA-303 Resin (80%)	190.0	25.00
Adjust speed to disperse pigments:		
Ti-Pure R-901 Titanium	300.0	9.04
Nopco NDW Defoamer	2.0	0.26
Slow to mixing speed and add:		
Ethylene Glycol	46.5	5.00
Nuodex PMA-18 Fungicide	3.0	0.30
Cellosize QP-4400 (3%) Thickener	2.0	–
Ucar 180 Vinyl Acetate/Acrylic Copolymer Resin (55%)	415.0	43.92
Super-Cobalt Drier	1.5	0.18
Ammonia (28%)	2.0	0.25

Visc. (KU) – Fresh	86
30 days	86
PVC (%)	16.3
Total Solids (% by wt.)	61.9
Wt. per gal (lb)	10.9
60° gloss reading of dry film	42

Chemacoil TA-303 = 40% of binder solids

Exterior Semigloss Silicone-Alkyd Copolymer

Formula	No. 1 (White)	No. 2 (Blue)	No. 3 (Navy Haze Gray)
Rutile Titanium Dioxide	26.6	15.69	16.3
Lamp Black	—	0.57	0.3
Ramapo Blue	—	0.86	—
Asbestine 325	11.6	11.92	16.2
3000 Mesh Mica	4.8	5.61	5.2
Soya Lecithin	1.0	1.02	1.0
Thixcin R	0.3	0.29	0.3
Silicone Alkyd (60% N.V.)	42.4	47.40	45.5
Naphtha Mineral Spirits	12.6	—	14.0
Mineral Spirits	—	15.65	—
Cobalt Octoate (6%)	0.3	0.28	0.3
Manganese Naphthenate (6%)	0.1	0.14	0.2
Calcium Naphthenate (5%)	0.2	0.28	0.3
Antiskinning Agent	0.1	0.08	0.1
Dow-Corning Paint Additive I	—	0.21	—
Paint Additive (Antiflooding, Floating, and Silking Agent)	—	—	0.3

White Polyvinyl Acetate Latex House Paint

(Oil Modified)

	lb/100 gal (approx.)
Water	155.2
Bentone LT Gellant	3.5

(Mix for 5–10 min at highest possible speed)

Victawet 35-B	5.4
Igepal CTA-639	1.7
Daxad 30	5.0
Colloid 677	1.0
Water	113.5

	lb/100 gal (approx.)
Titanox RA-NC 1	177.8
Oncor 45X	119.9
Foam A/Baryta White	39.2
Phenyl Mercuric Borate	0.8

(Disperse 20 min at high speed)

Ethylene Glycol		24.9
Hexylene Glycol		3.7
Colloid 677		1.0
Resyn 2243	Add slowly till uniform	164.4
DCO Castung 235		139.5
Lead Naphthenate (24%) } Premix		5.1
Cobalt Naphthenate (6%)		2.2
Water		139.8
Wt./gal (lb)		11.0
PVC (%)		26.3

Brookfield Visc. (cps)	
10 rpm	5200
50 rpm	3400
50 rpm	2080
100 rpm	1440
Brushometer (poiss)	0.99
Cone & Plate (dyne/cm^2)	12.0

Stormer Visc.	
(g)	230
(KU)	86

Exterior White Paint

FORMULA NO. 1

(Polyvinyl Acetate)

	lb	gal
Water	115.4	13.87
Tamol 850	5.0	0.51
Potassium Tripolyphosphate	2.0	0.10

	lb	gal
Propylene Glycol	25.0	2.90
Colloid 581-B Antifoamer	2.0	0.30
Ti-Pure R-902 (Rutile)	236.2	7.14
Ti-Pure LW (Anatase)	19.2	0.59
Methocel 90HG, DG4000		
(2% sol'n.)	75.0	8.97
Asbestine 325 Talc	144.9	6.17
Celite 281	25.0	1.30
Mica (325-mesh, water ground)	25.0	1.06
Metasol 57 Preservative (bulk)	1.8	0.11

Disperse with high-speed disk dispenser, then add:

	lb	gal
Elvace PB3-1952	397.7	44.69
Methocel 90HG, DG4000		
(2% sol'n.)	90.0	10.77
Water	12.7	1.53

PVC (%)	40.6
Solids (% by Wt.)	57.7
Solids (% vol.)	40.1
Theor. 1 mil coverage/gal	643 ft^2

No. 2

(Polyvinyl Acetate)

	lb	gal
Dispersion:		
Water	100.0	12.00
Daxad 30 (25%)	8.4	0.87
Tetrapotassium Pyrophosphate	1.0	0.05
Igepal CTA-639	1.5	0.17
Polyglycol P-1200	2.0	0.29
Methocel Sol'n. (3%) (65 HG,		
4000 DG)	70.0	8.48
Titanox RANC	225.0	6.43
Ti-Pure FF	25.0	0.77
Nytal 300	50.0	2.10
Mica 325 Mesh	30.0	1.30

	lb	gal
Water	20.0	2.42
Foamicide 581-B	1.0	0.15

Reduction:

	lb	gal
Carbitol Acetate ⎱ Premix	20.0	2.42
Water ⎰	20.0	2.42
Methocel Sol'n. (3%) (65 HG, 4000 DG)	68.5	8.23
Everflex BG	425.0	47.25
Phenyl Mercuric Acetate (18%)	5.0	0.45
Ethylene Glycol	40.0	4.31
Additive Sol'n.*	98.18	12.50

PVC (%)	32
Total Solids (%)	53
Total Solids (% vol.)	40.6
Visc. (KU)	82–86

*Additive Sol'n.:

	lb	gal
Castung 235	92.70	11.96
Phenyl Mercuric Acetate (18%)	2.78	0.25
Naphthenate Driers (24% Lead)	3.5	0.36
Naphthenate Driers (6% Cobalt)	1.5	0.19

No. 3

(Polyvinyl Acetate)

	lb	gal
Premix:		
Water	203.0	24.36
Strodex PK-90 Dispersing Agent	2.0	0.21
Triton X-100 Surfactant	2.0	0.23
Potassium Tripolyphosphate	1.0	—
Ammonia (28%)	3.0	0.38

Increase speed and add slowly:

	lb	gal
Chemacoil TA-100	120.0	15.37

	lb	gal
Adjust speed to disperse pigments:		
Ti-Pure R-901 Titanium Dioxide	175.0	5.11
Titanox A-168-LO Titanium Dioxide	25.0	0.77
Nytal 300 Talc	150.0	6.31
Mica (325-mesh, water-ground)	30.0	1.26
Ethylene Glycol	18.5	2.00
Nuodex PMA-18 Fungicide	3.0	0.26
Colloid 581-B Defoamer	1.0	0.15
Slow to mixing speed:		
Cellosize QP-15000 (2%) Thickener	183.0	21.94
Resyn 1255 Polyvinyl Acetate		
Emulsion	215.0	23.62
Super-Cobalt Drier	1.0	0.13

Visc. (KU) –	Fresh		82
	Overnight		82
	15 months		95
PVC (%)			33.1

Chemacoil TA-100 = 50% of binder solids

No. 4

(Polyvinyl Acetate)

	lb	gal
Premix:		
Water	150.0	18.00
Daxad 11 Dispersing Agent	1.5	0.12
Potassium Tripolyphosphate	4.0	—
Triton X-100 Surfactant	2.3	0.26
Increase speed and add slowly:		
Chemacoil TA-100	150.0	19.23
Adjust speed to disperse pigments:		
Ti-Pure R-901 Titanium Dioxide	200.0	5.84
Duramite Calcium Carbonate	225.0	10.04

	lb	gal
Celite 281 Diatomaceous Silica	50.0	2.60
Mica (325-mesh, water-ground)	25.0	1.06
Colloid 581-B Defoamer	2.0	0.30
Ethylene Glycol	18.5	2.00
Nuodex PMA-18 Fungicide	3.0	0.26

Slow to mixing speed:

	lb	gal
Cellosize QP-15000 (2%) Thickener	122.0	14.64
Resyn 2243 Polyvinyl Acetate Emulsion	248.0	27.25
Ammonia (28%)	4.0	0.52
Super-Cobalt Drier	1.0	0.13

Visc. (KU) – Fresh		84
Overnight		84
10 months		86
PVC (%)		37

Chemacoil TA-100 = 50% of binder solids

Exterior White Primer

(Polyvinyl Acetate)

	lb	gal

Premix:

	lb	gal
Water	125.0	15.00
Daxad 30 Dispersing Agent	10.0	1.18
Tergitol NPX Surfactant	4.6	0.52
Victawet 35-B Surfactant	3.0	0.31

Increase speed and add slowly:

	lb	gal
Chemacoil TA-100	78.0	10.00

Adjust speed to disperse pigments:

	lb	gal
202 Basic Silicate White Lead	125.0	2.34
Ti-Pure R-901 Titanium Dioxide	200.0	5.84
Titanox A-168-LO Titanium Dioxide	20.0	0.62
Nytal 300 Talc	50.0	2.10
Mica (325-mesh, Water-Ground)	25.0	1.06

	lb	gal
Ethylene Glycol	40.0	4.32
Polyglycol P-1200 Polypropylene Glycol	3.0	0.36
Nuodex PMA-18 Fungicide	3.0	0.26
Colloid 581-B Defoamer	2.0	0.30

Slow to mixing speed:

	lb	gal
Cellosize QP-15000 (2%) Thickener	186.0	16.32
Everflex BG Vinyl Acetate Copolymer		
Emulsion (52%)	355.0	39.88
Super-Cobalt Drier	0.7	0.09

Visc. — Fresh	82	KU
Overnight	82	KU
6 months	87	KU
PVC (%)	29.2	

Chemacoil TA-100 = 30% of binder solids

Self-Priming Exterior White Paint

(Vinyl Acetate)

	lb	gal
Premix:

	lb	gal
Water	120.0	14.40
Makon 10 Surfactant	8.0	0.90
Triton X-100 Surfactant	4.0	0.45
Clearate WD Lecithin	8.0	0.91

Increase speed and add slowly:

	lb	gal
Chemacoil TA-100	78.0	10.00

Adjust speed to disperse pigments:

	lb	gal
202 Basic Silicate White Lead	125.0	2.34
Ti-Pure R-901 Titanium Dioxide	200.0	5.84
Titanox A-168-LO Titanium Dioxide	20.0	0.62
Nytal 300 Talc	50.0	2.10
Mica (325-mesh, Water-Ground)	25.0	1.06
Polyglycol P-1200 Polypropylene Glycol	3.0	0.36

	lb	gal
Ethylene Glycol	40.0	4.32
Nuodex PMA-18 Fungicide	3.0	0.27

Slow to mixing speed:

	lb	gal
Cellosize QP-15000 (2%) Thickener	119.0	14.28
Polyco 804-PL Vinyl Acetate Copolymer		
Emulsion (55%)	355.0	39.01
Super-Cobalt Drier	1.0	0.13
Water	18.0	2.16

Visc. — Fresh	87	KU
Overnight	90	KU
12 months	95	KU
PVC (%)	29.9	

Chemacoil TA-100 = 30% of binder solids

Polyvinyl Acetate Latex Paint

	lb
Water	185.0
Renex 690 Wetting Agent	3.0
G-3300 Dispersing Agent	3.0
Colloid 581B Defoamer	2.0
TiPure R-901 Titanium Dioxide	200.0
Satintone #1 Calcined Clay	50.0
Celite 281 Diatomaceous Silica	30.0
Asbestine -3X Talc	138.0

Exterior Copolymer Polyvinyl Acetate Emulsion Paints

FORMULA	No. 1 (White)		No. 2 (Tint Base)	
	lb	gal	lb	gal
TiPure R-900	200	5.9	150	4.4
Drikalite	350	15.6	200	8.9
Daxad 30	5	0.5	5	0.5
Colloid 606	2	0.3	2	0.3

	No. 1 (cont'd) (White)		No. 2 (cont'd) (Tint Base)	
	lb	gal	lb	gal
Ethylene Glycol	28	3.0	28	3.0
Troysan PMA-30	4	0.4	4	0.4
Igepal CO-610	2	0.2	2	0.2
Natrosol 250 HR (3%)	58	7.0	75	9.0
Water	184	22.1	177	21.3
Flexbond 800 (52 NV)	405	45.0	467	52.0
PVC (%)	49.8		34.5	
Visc. (KU)	80–85		80–85	
Paint NV (% by vol.)	43.9		37.6	
Paint NV (% by Wt.)	61.5		53.7	
Vehicle NV (% by Wt.)	30.6		32.3	

	No. 3 (White)		No. 4 (Tint Base)	
	lb	gal	lb	gal
TiPure R-900	200	5.9	150	4.4
Drikalite	350	15.6	200	8.9
Daxad 30	5	0.5	5	0.5
Colloid 606	2	0.3	2	0.3
Ethylene Glycol	28	3.0	28	3.0
Troysan PMA-30	4	0.4	4	0.4
Igepal CO-519	2	0.2	2	0.2
Natrosol 250 HR (3%)	58	7.0	67	8.0
Water	184	22.1	186	22.3
Everflex G (52 NV)	405	45.0	463	52.0
PVC (%)	49.8		34.9	
Visc. (KU)	80–85		80–85	
Paint NV (% by vol.)	43.9		38.1	
Paint NV (% by Wt.)	61.5		53.6	
Vehicle NV (% by Wt.)	30.6		32.0	

	No. 5 (White) lb	gal	No. 6 (Tint Base) lb	gal
Ti-Pure R-900	200	5.9	150	4.4
Drikalite	350	15.6	200	8.9
Daxad 30	5	0.5	5	0.5
Colloid 606	2	0.3	2	0.3
Ethylene Glycol	28	3.0	28	3.0
Troysan PMA-30	4	0.5	4	0.4
Igepal CO-610	2	0.2	2	0.2
Natrosol 250 HR (3%)	33	4.0	67	8.0
Water	234	28.0	211	25.3
Resyn 1251 (55 NV)	382	42.0	445	49.0
PVC (%)	50.0		34.9	
Visc. (KU)	80–85		80–85	
Paint NV (% by vol.)	43.0		38.1	
Paint NV (% by Wt.)	61.8		53.6	
Vehicle NV (% by Wt.)	31.3		32.5	

Exterior Acrylic Emulsion Paints

Formula	No. 1 (White) lb	gal	No. 2 (Tint Base) lb	gal
TiPure R-900	200	5.9	150	4.4
Drikalite	350	15.6	200	8.9
Daxad 30	5	0.5	5	0.5
Colloid 606	2	0.3	2	0.3
Ethylene Glycol	28	3.0	28	3.0
Troysan PMA-30	4	0.4	4	0.4
Igepal CO-610	2	0.2	2	0.2
Natrosol 250 HR (3%)	58	7.0	67	8.0
Water	159	19.1	152	18.3
Polyco 2719 (46.5 NV)	415	48.0	484	56.0

	No. 1 (cont'd) (White)		No. 2 (cont'd) (Tint Base)	
	lb	gal	lb	gal
PVC (%)		50.0		34.7
Visc. (KU)		80–85		80–85
Paint NV (% by vol.)		43.0		38.3
Paint NV (% by Wt.)		61.5		52.8
Vehicle NV (% by Wt.)		29.1		30.4

Exterior Ethylene – Vinyl Acetate Emulsion Paints

FORMULA	No. 1 (White)		No. 2 (White)	
	lb	gal	lb	gal
TiPure R-900	200	5.9	200	5.9
Drikalite	350	15.6	350	15.6
Daxad 30	5	0.5	5	0.5
Colloid 606	2	0.3	2	0.3
Ethylene Glycol	28	3.0	28	3.0
Troysan PMA-30	4	0.4	4	0.4
Igepal CO-610	2	0.2	2	0.2
Natrosol 250 HR (3%)	58	7.0	58	7.0
Water	209	25.1	209	25.1
Aircoflex 500 (55 NV)	382	42.0	–	–
Elvace 1952 (55 NV)	–	–	373	42.0
PVC (%)		50.0		50.0
Visc. (KU)		80		80
Paint NV (% by vol.)		43.2		43.2
Paint NV /% by Wt.)		61.8		61.2
Vehicle NV (% by Wt.)		30.5		30.1

| | No. 3 | | No. 4 | |
| | (Tint Base) | | (Tint Base) | |
	lb	gal	lb	gal
TiPure R-900	150	4.4	150	4.4
Drikalite	200	8.9	200	8.9
Daxad 30	5	0.5	5	0.5
Colloid 606	2	0.3	2	0.3
Ethylene Glycol	28	3.0	28	3.0
Troysan PMA-30	4	0.4	4	0.4
Igepal CO-610	2	0.2	2	0.2
Natrosol 250 HR (3%)	84	10.0	91	10.9
Water	202	24.3	195	23.4
Aircoflex 500 (55 NV)	437	48.0	—	—
Elvace 1952 (55 NV)	—	—	428	48.0
PVC (%)	35.0		35.0	
Visc. (KU)	80		80	
Paint NV (% by vol.)	38.3		38.3	
Paint NV (% by Wt.)	53.3		53.2	
Vehicle NV (% by Wt.)	31.8		30.1	

High-Gloss White Exterior

(Polyester)

	lb	gal
Pigment Dispersion:		
Rutile Titanium Dioxide	300	8.60
Sag and Suspension Control Agent	15	2.00
Flow Control Agent	3	0.37
Aroflint 303-X-90	200	24.40
Aroflint 303-X-90	100	12.20
Solvent 150	18	2.43
Clear Activatior:		
Aroflint 202-XA1-60	364	35.40

	lb	gal
Cellulose Acetate Butyrate		
Sol'n. (15%)	42	5.00
Ethylene Glycol Ethyl Ether Acetate	78	9.60

PVC (%)	15.0
Total Solids, Wt. (%)	71.2
Total Solids, vol (%)	58.9
Vehicle Nonvolatile (%)	60.8
Visc. (KU)	65–75
Ratio components **Aroflint 202/303** solids	45/55

Blend the formula one-to-one by volume on the job.

Semigloss White Exterior

(Polyester)

	lb	gal
Pigment Dispersion:		
Rutile Titanium Dioxide	300	8.60
Micronized Magnesium Silicate	100	4.22
Flow Control Agent	3	0.37
Suspension and Antisag Agent	15	2.00
Aroflint 303-X-90	200	24.40
Aroflint 303-X-80	43	5.25
Solvent 150	38	5.16
Clear Activator:		
Aroflint 202-XA1-60	299	29.00
Cellulose Acetate Butyrate Sol'n. (15%)	42	5.00
Ethylene Glycol Ethyl Ether Acetate	88	11.00
Solvent 150	37	5.00

60° Gloss	70.80
PVC (%)	24.3
Total Solids, Wt. (%)	70.1
Total Solids, vol. (%)	53.4
Vehicle Nonvolatile (%)	54.0
Visc. (KU)	60–70
Ration components **Aroflint 202/303** solids	45/55

Blend formula one-to-one by volume on the job.

High Gloss Green Exterior Paint

(Polyester)

	lb	gal
Pigment Dispersion:		
Rutile Titanium Dioxide	180	5.15
Phthalocyanine Green	20	1.34
Sag and Suspension Control Agent	15	2.00
Flow Control Agent	3	0.37
Aroflint 303-X-90	170	20.80
Aroflint 303-X-90	130	15.80
Solvent 150	34	4.54
Clear Activator:		
Aroflint 202-XA1-60	364	35.40
Cellulose Acetate Butyrate Sol'n. (15%)	42	5.00
Ethylene Glycol Ethyl Ether Acetate	78	9.60
PVC (%)	11.7	
Total Solids (% Wt.)	67.5	
Total Solids (% vol.)	56.8	
Vehicle Nonvolatile (%)	59.8	
Visc. (KU)	60–70	
Ratio components **Aroflint 202/303** solids	45/55	

Blend formula one-to-one by volume on the job.

Low-Gloss White Exterior

(Polyester)

	lb	gal
Pigment Dispersion:		
Rutile Titanium Dioxide	250	7.15
Magnesium Silicate	350	14.75
Phenyl Mercury Preservative	3	0.20
Sag and Suspension Control Agent	4	0.50
Arofiint 303-X-90	217	26.50
150 Solvent	100	13.50
Solvent 150	30	4.07

	lb	gal
Clear Activator:		
Aroflint 202-XA1-60	268	25.90
Ethylene Glycol Ethyl Ether Acetate	60	7.43

60° Gloss	10
PVC (%)	38.4
Total Solids, Wt. (%)	74.6
Total Solids (vol.)	57.8
Vehicle Nonvolatile (%)	52.6
Visc. (KU)	65–75
Ratio components **Aroflint 202/303** solids	45/55

Blend formula two-to-one by volume on the job.

Mildew-Resistant Paint

(Tri-Ester)

	lb	gal
Ti-Pure R-900 Titanium Dioxide	200	5.72
Busan 11-MI Preservative	200	7.01
Nytal 300 Talc	100	4.20
Pentachlorophenol	6	0.77
Chemacoil TA303 Resin (80%)	456	60.00
Thixcin 25C	4	0.45
Methyl Ethyl Ketoxime	2	0.20
Cobalt Naphthenate (6%)	0.5	0.06
Lead Naphthenate (24%)	5	0.50
Mineral Spirits	150	23.09

lb/gal	11.1
PVC (%)	27.1
Solids (% by Wt.)	77.0
Visc. — Fresh (KU)	82
Overnight	82

Architectural Alkyd White Enamel

FORMULA NO. 1

	lb	gal
Rutile Titanium Dioxide	293	8.37
Kadox #15 Zinc Oxide	33	0.7
Varkyd 505-70	219	27.4

Grind on roller mill and add:

	lb	gal
Varkyd 505-70	350	43.7
Mineral Spirits	152	23.0
Cobalt Naphthenate (6%)	2.6	0.32
Calcium Naphthenate (4%)	4.25	0.55
Antiskinning Agent	1.0	0.1

Yield (gal)	101.5
Gal Wt. (lb)	10.3
PVC (%)	16.7
Gloss	Excellent
Brushing	Good
Leveling	Good
Stay-Put	Good

NO. 2

	lb	gal
Roller mill grind:		
Titanium Dioxide	278.0	8.40
Zinc Oxide	14.7	0.30
Duraplex D-65A (70% solids)	195.2	24.40
Mix with:		
Duraplex D-65A (70% solids)	315.8	39.48
Mineral Thinner	170.1	26.37
Cobalt Naphthenate (6%)	2.8	0.35
Calcium Naphthenate (6%)	5.6	0.70

Wt./gal (lb)		9.8
Total Solids (%)		66.5
Pigment (%)	45.0	
Binder (%)	55.0	
Drier (metal on resin solids)		
Cobalt Naphthenate (%)	0.05	
Calcium Naphthenate (%)	0.10	

Alkyd Architectural Gloss Enamel

FORMULA NO. 1

	lb
Rutile Titanium Dioxide	220
Rutile Titanium Calcium	220
Alkyd Vehicle (70% N.V.)	475
Mineral Spirits	175
Zinc Naphthenate (24%)	3
Cobalt Naphthenate (6%)	2
Antiskinning Agent	1

No. 2

	lb	gal
RA-50 or Equivalent Titanium Dioxide	250	7.2
Kadox 515 Zinc Oxide	25	0.5
Mineral Spirits	40	6.0
Varkyd 700-70	328	42.0

2 passes on roller mill thin with the following items:

Varkyd 700-70	172	22.0
Mineral Spirits	159	24.0
Cobalt Naphthenate (6%)	3	0.4
Calcium Naphthenate (4%)	5	0.6
Volatile ASA Exkin #2	1	0.1

PVC (%)	16.7
Vehicle Nonvolatile (Wt.)	50.0
Visc. (KU)	80–85
Brushing	Very Good
Leveling	Excellent

Sagging Resistance	Very Good
Settling	None
Skinning	None
Gloss	Excellent
Dry-Set-to-Touch	1–2 hr
Hard	5–6 hr

Alkyd Green Trim and Trellis Enamel

	FORMULA No. 1	No. 2
	lb	gal
Rutile Titanium Dioxide	100.0	2.92
Pure Chrome Green	100.0	2.86
Litharge	1.0	0.01
Chlorinated Paraffin	52.0	3.91
Syntex 77	589.0	77.00
Post-4	8.0	0.92
Mineral Spirits	50.0	7.58
Lead Naphthenate (24%)	4.3	0.44
Cobalt Naphthenate (6%)	2.3	0.28
Manganese Naphthenate (6%)	2.3	0.28
Antiskinning Agent	1.0	0.13

Performance:	Control (No Additive)	Post-4 (Added in the Grind)
Level lb/PHG	0	8
Visc. (KU)		
(1 day)	91	93
(1 week)	97	97
Sag Rating (Leneta)		
(1 day)	Passed 3	Passed 5
(1 week)	Passed 3	Passed 6+

Fast-Drying, High-Gloss Enamel — Interior and Exterior

(Alkyd)

	lb	gal
RA-50 Titanium Dioxide	250	7.2
Atomite Calcium Carbonate	50	2.2
Soya Lecithin	3	.4
Mineral Spirits	20	3.0
Varkyd 552-50	138	18.0

2 passes on a roller mill and thin with the following:

	lb	gal
Varkyd 552 (50%)	444	58.0
Mineral Spirits	86	13.0
Cobalt Naphthenate (6%)	2	0.3
Calcium Naphthenate (4%)	3.5	0.4
Lead Naphthenate (24%)	4	0.4
ASA Exkin #2	0.5	—

PVC (%)	22.0
Vehicle Nonvolatile (Wt.) (%)	42.6
Total NV by Wt. (%)	59.4
Wt./gal (lb)	9.72
Brushing	Excellent
Leveling	Excellent
Sagging Resistance	Very Good
Settling	None
Skinning	None
Gloss	Excellent
Dry — Set-to-Touch	1 hr
Hard	2–4 hr

Exterior Silicone Alkyd Copolymer Gloss Enamel

	lb	gal
Varkyd 333-60	216	27.0
Mineral Spirits	19.5	3.0
Toluol	47	6.5
Cellosolve	23	3.0
Soya Lecithin	3	.4
MPA-60X	8	.9
Titanium Dioxide RANC	290	8.5

	lb	gal
Pebble or sand mill:		
Varkyd 333-60	400	50.0
Cobalt Naphthenate (6%)	2.5	.3
Calcium Naphthenate (4%)	4.5	.5
Lead Naphthenate (24%)	5	.5
Vol. Antiskinning Agent	1	.1

PVC (%)	17.7	
Vehicle Nonvolatile (Wt.) on Enamel (%)	36.3	36 min
Total NV by Wt. (%)	65.3	65 min
Pigm. Solids by Wt. of Paint (%)	28.3	27–31
Wt./gal (lb)	10.11	
Visc. (KU)	68–71	67–77
Dilution Stability	Passes	
Storage Stability (Part Cont)	Passes	
Brushing	Passes	
Leveling	Passes	
Sagging Resistances	Passes	
Settling	Passes	
Skinning	Passes	
Gloss	93.5	87 min
Water Resistance	Passes	
Dry — Set-to-Touch	1/2 hr	2 hr max
Hard	4–6 hr	8 max (hr)
Reflectance	89	87
Contrast Ratio	0.91	0.90 min
Recoatability	Passes	
Weatherability	Passes	300 hr in weatherometer
Hydrocarbon Resistance	Passes	

White Semigloss Enamel

(Oil)

	lb	gal
Pigment Dispersion:		
Aroflint 252 Component	270.0	28.80

	lb	gal
Ethylene Glycol Monethyl Ether Acetate	16.2	2.00
Rutile Titanium Dioxide	300.0	8.58
Calcium Carbonate	200.0	8.90
Magnesium Silicate	175.0	7.35
Thixatrol ST	1.0	0.16

Grind on high speed disperser to 140°F:

	lb	gal
Exempt Mineral Spirits	31.6	4.88
Aroflint 252 Component	57.2	6.00

Clear Activator:

	lb	gal
Aroflint 606 Component	160.0	19.05
Exempt Minteral Spirits	71.5	11.00
Ethylene Glycol Monoethyl Ether Acetate	26.6	3.28

Visc. (KU)	67–72
lb/gal	13.1
60°Gloss	40–50
Total Solids (% by Wt.)	78.9
Total Solids (% by vol.)	59.5
PVC (%)	41.8

Isophthalic Enamelized House Paint

(Alkyd)

	lb	gal
Titanox RA-50	275	7.85
Vicron	100	4.45
Thixcin R	4	.5
Varkyd 701-70	385	49.0
Mineral Spirits	20	3.0

Roller mill grind:

	lb	gal
Varkydol 210-100	121	15.0
Mineral Spirits	152	23.0
Cobalt Naphthenate (6%)	2	.3

	lb	gal
Lead Naphthenate (24%)	3	.3
Volatile ASA	1	.1
Super-Ad-It	7	1.0

PVC (%)	20.8
Vehicle Nonvolatile	58.0
Total NV (% by Wt.)	71.8
Wt./gal (lb)	10.25
Visc. (KU)	75–80
Brushing	Excellent
Leveling	Very Good
Sagging Resistance	Excellent
Settling	None
Skinning	None
Gloss	Excellent
Dry — Set-to-Touch	3–4 hr
Hard	8–10 hr

Exterior Semigloss Latex Trim Paint

	lb	gal
Premix:		
Water	108.3	13.00
Daxad 30 Dispersant	8.0	0.94
Tergitel NPX Surfactant	5.0	0.56
Victawet 35-B Surfactant	3.0	0.31
Increase speed and add slowly:		
Chemacoil TA-100	78.0	10.00
Duramac 2438 Alkyd Resin	116.0	14.53
Adjust speed to disperse pigments:		
Ti-Pure R-901 Titanium Dioxide	300.0	8.76
Cellosize QP-15000 (dry) Thickener	0.5	—
Polyglycol P-1200 Polypropylene Glycol	6.0	0.71
Ethylene Glycol	37.0	4.00

	lb	gal
Nuodex PMA-18 Fungicide	4.5	0.41
Colloid 581-B Defoamer	1.0	0.15

Slow to mixing speed:

	lb	gal
Everflex BG Vinyl Acetate Copolymer Emulsion (52%)	437.0	49.10
Super-Cobalt Drier	1.0	0.13

Visc. (KU) — Fresh	72
Overnight	72
36 months	83
PVC (%)	12.0

Chemacoil TA-100 = 20% of binder solids

Exterior White Enamel

(Tri-Amino)

	lb	gal
Ti-Pure R-900 Titanium Dioxide	300.0	8.58
Nytal 300 Talc	200.0	8.40
Chemacoil TA-303 Resin (80%)	432.0	56.44
Mineral Spirits	172.0	26.42
Thixcin 25C Thickener	6.0	0.92
Cobalt Drier (6%)	0.5	0.06
Lead Drier (24%)	5.0	0.52
Exkin #2 Antiskinning Agent	2.0	0.26

Visc. (KU — Fresh	79
PVC (%)	28.3
Total Solids (% by Wt.)	75.4
Wt./gal (lb)	11.0

This exterior white enamel house paint provides better adhesion to aluminum than do white enamels formulated using oil-modified urethanes or medium oil soya alkyds.

Yellow Trim Enamel

(Tri-Ester)

	lb	gal
Chrome Yellow Medium	225.0	4.59
Ferrite Yellow	22.5	0.68
Bentone 38 Gel (10% in mineral spirits)	9.0	1.29
Chemacoil TA-303 Resin (80%)	427.1	56.00
Mineral Spirits	221.0	34.00
Cobalt Drier (6%)	2.0	0.25
Lead Drier (24%)	4.0	0.42
Zinc Drier (8%)	2.0	0.25
Antiskinning Agent	2.0	0.28
Maxvar 2503 Additive (55%)	25.6	3.50

Visc. — Fresh (KU)	80
PVC (%)	10.7
Vehicle Nonvolatile (% by Wt.)	52.3
Total Solids (% by vol.)	64.1
Wt./gal (lb)	9.2

This formulation has undergone several years' exposure on buildings in Wisconsin and in Indiana, with excellent weathering characteristics. It showed outstanding one-coat hold out and gloss retention over old weathered siding.

Exterior Vinyl — Acrylic Topcoat Enamel

Formula	No. 1	No. 2
	lb	to make 100 gal (approx.)
Water	85.7	85.7
Surfactant[1]	3.0	3.0
Defoamer[2]	1.0	1.0
Propylene Glycol	25.0	25.0
Thickener Solution[3]	95.0	95.0
Titanium Dioxide[4]	250.0	250.0
Calcium Carbonate[5]	205.0	205.0

	No. 1	No. 2
		to make 100 gal
	lb	(approx.)
Let down with water to grinding viscosity.	75.0	25.0
Grind then add:		
Vinyl-Acrylic Latex (55%)[6]	205.0	—
Defoamer[2]	1.0	1.0
Preservative[7]	2.0	2.0
NH_4OH (28%)*	5.4	5.4
Water*	50.0	70.0
Resin WS-220T*	174.0	174.0
Acrylic Latex (46%)[8]	—	244.0
Water	84.0	75.0
Solids (%)	54	54
PVC (%)	39	39

*Premix NH_4OH, water, and **WS-220T** before adding to paint.

[1] **Triton X-100**
[2] **Foamaster**
[3] **250HR (2½%)**
[4] **TiPure R-900**
[5] **Duramite**
[6] **Resyn 25-2243**
[7] **Super-Ad-It**
[8] **Rhoplex AC-35**

Latex Trim Paint

	to make 93 gal (approx.)
Resin WS-220T (65% NVM)	206
Propylene Glycol	25
Surfactant[1]	3
Ammonium Hydroxide (28%)	5.0
Titanium Dioxide[2]	275
Defoamer[3]	2

Add the above at low speed in the order shown then grind at high speed. Add the following at low speed:

	to make 93 gal (approx.)
Water	133
Defoamer[3]	2
Preservative[4]	2
Vinyl-Acrylic Latex (55% NVM, pH 7.0)[5]	244
Water	129
Water or Thickener[6] to final viscosity.	29
Adjust pH with amine to	8–9
Solids (%)	51.5
PVC (%)	21
Stormer Visc. (KU)	70–75

---

[1] Triton X-100
[2] Ti-Pure R-900
[3] Foamaster
[4] Super-Ad-It
[5] Resyn 25-2243 Adjust pH to 7.0 with amine before using.
[6] ASE-60 (5%) (Thickened with TEA or NH_4OH)

Latex Semigloss Enamel

(Vinyl Acrylic)

	lb	gal
Ethylene Glycol	100.0	10.30
KTPP	1.5	—
Tamol 731-25	7.5	0.82
Defoamer 2153	3.0	0.41
Rutile Titanium Dioxide	250.0	7.35
Sparmite	50.0	1.36
Kaopaque 6	25.0	1.13
Defoamer 2153	1.0	0.13
Water	152.0	18.24
Carbitol	10.0	1.21
NH_4OH Sol'n. (9:1 water: conc. NH_4OH)	15.0	1.80
Triton X-102	4.0	0.50
Varaqua 939	540.0	60.00

PVC (%)	23.8
Vehicle Nonvolatile (Wt.) (%)	36.5
Total NV by Wt. (%)	54.2
Wt./gal (lb)	11.2
Visc. (KU — sheared)*	70–80
Gloss photovolt — 60°	45–55

---

*Viscosity should not be adjusted until the day after the let down is accomplished. Dilute NH_4OH solutions under agitation should be used to raise viscosity. Additional adjustment should not be attempted until 4 hr after the initial adjustment ammonia addition.

Laboratory tests indicate excellent acceptance of Cal. Ink 8800 series colorants. Include 1 lb of preservative.

White Gloss Enamel

(Acrylic)

	lb	gal
Water	83.3	10.0
Ammonia (Conc)	2.0	0.3
TKPP	1.0	—
Phtalopal SEB (30%)	19.5	2.2
Latekoll LR 8234 (45%)	18.0	2.1
PMA-30	0.3	—
Ethylene Glycol	74.0	8.0
Texanol	25.0	3.2
Mineral Spirits	19.5	3.0
Titanox CL	275.0	7.9

Mix for 10 min on high speed disperser and let down.

Deefo 97-2	1.5	0.2
Ammonia (Conc.)	2.0	0.2
Acronal 290 D	500.0	58.1
Water	50.0	6.0

Add ingredients in order given and heat to 120°F under agitation.

Phtalopal SEB (30%)	
Water	63.0
Ammonium Hydroxide (28%)	7

Phtalopal SEB Resin 30

PVC (%)	21.2
Nonvolatile (% by vol.)	36.7
Nonvolatile %	50.4
lb/gal	10.58
Visc. (KU)	80±5
pH	75±5
60° Gloss	8.8–9.3

Vinyl/Acrylic Semigloss Latex Enamel

	lb	gal
Premix:		
Propylene Glycol	77.1	9.00
AMP-95	4.0	0.51
Potassium Tripolyphosphate	1.0	–
Nopco NDW Defoamer	2.0	0.26
Disperse pigments:		
Titanox 2020 Titanium Dioxide	250.0	7.32
Barytes	75.0	2.04
Grind; then add:		
Propylene Glycol	52.0	6.00
Cellosize QP-4400 (2½%) Thickener	41.8	5.00
Poly-Tex 6400 Vinyl Acetate/Acrylic Emulsion (50%)	522.0	58.00
Water	24.9	2.99
Cellosize QP-4400 (2½%) ⎫	25.2	3.02
Dowicil 75 Preservative ⎬ Premix	1.0	0.10
Igepal CO-630 Surfactant ⎪	4.4	0.50
Texanol Coalescing Aid ⎭	10.0	1.25
Cellosize QP-4400 (2½%) Thickener	32.3	3.88
Nopco NDW Defoamer	1.0	0.13

Visc. (KU) at 77 F — Fresh	65 KU
— Overnight	67 KU
— 1 week	67 KU
— 1 month	70 KU
— 1 month at 125 F	70 KU

pH — Fresh	9.5
1 month at 125 F	8.4
PVC (%)	25.8
Total Solids (% by Wt.)	52.2
60° Gloss reading of film	45.0

White Exterior Gloss Trim

(Acrylic)

Propylene Glycol	78
Tamol 731	16
Triton CF-10	1
Dapro 881-S Defoamer	1
Water	33
Du Pont R-900	283

Mix on high speed mill and add:

Texanol	4
ParCryl 400	505
Dapro 881-S	1
Cosan 171-S	9
Ammonium Hydroxide	2
Propylene Glycol	42
Water	100
Acrysol G-110 (as needed)	10
PVC (%)	23
Solids (%)	50
Visc. (KU)	68–72
pH	9.5

Acrylic Emulsion Gloss Latex Enamel

	lb	gal
Premix:		
Ethylene Glycol	102.5	11.04
AMP-95	4.0	0.51
Potassium Tripolyphosphate	2.0	—
Nopco NDW Defoamer	2.0	0.26

	lb	gal
Disperse pigments:		
Titanox 2020 Titanium Dioxide	275.0	8.02
Grind; then add:		
Propylen Glycol	50.0	5.81
Rhoplex AC-490 Acrylic Emulsion (46.5%)	590.5	66.61
Nopco NDW Defoamer	4.0	0.52
Tris Nitro (50%) Preservative	1.0	0.12
Butyl **Cellosolve**	27.5	3.51
Triton GR-7 Surfactant	2.0	0.23
Cellosize QP-4400 (2½%) Thickener	28.1	3.37

(Premix: Tris Nitro through Cellosize QP-4400)

Visc. (KU) at 77 F — Fresh	64
— Overnight	68
— 1 week	69
— 1 month	77
— 1 month at 125 F	87
pH — Fresh	9.4
1 month at 125 F	9.1
PVC (%)	20.3
Total Solids (% by Wt.)	50.5
60° Gloss reading of film	73.3

Low-Cost Shingle and Shake Paint

(Linseed Oil)

	lb	gal
Rutile Titanium Dioxide (**Titanox RA-50**)	150.0	4.29
Calcium Carbonate (**Snowflake**)	400.0	17.80
Celite 281	75.0	3.91
Aluminum Stearate (#27 Harshaw Chem.)	7.5	0.94
Carbon Black (Black Pearls #81)	2.0	0.13
Cykelin 70	320.0	41.50

	lb	gal
Kelecin F	6.0	0.75
Mineral Spirits	200.0	30.60
Cobalt Naphthenate (6%) – 0.15%	0.5	0.06
Manganese Naphthenate (6%) – 0.015%	0.5	0.06
Lead Naphthenate (24%) – 0.6%	5.0	0.50
Calcium Naphthenate – (4%) – 0.1%	6.0	0.70
Visc. (KU)	80	
PVC (%)	50.1	
Vehicle Solids (%)	43.1	
Total Solids (%)	74.1	
Wt./gal (lb)	11.55	

Shake and Shingle Paint

(Alkyd)

	lb	gal
Titanium Dioxide	200.0	
Camel Carb or equal	275.0	
Celite 281	25.0	
X/Y/Z (Troy) or equal	3.0	
Soya Lecithin	1.0	
FAGL 60 Long Oil Alkyd 60% N.V.	55.0	7
FAFLM	435.0	61
Slow Solvent	66.0	10
Cobalt Drier (6%)	1.6	
Lead (24%)	3.9	
Antiskinning Agent	1.0	
P.M.O. 30	8.8	
Yield (gal)	100	
PVC (%)	53	
N.V. on Vehicle (%)	30	

Alkyd Modified Latex House Paint

	lb	gal
Horse Head R-760	210	6.09
Horse Head A-440	20	0.62

	lb	gal
Horse Head XX-503	100	2.14
Asbestine 325	108	4.56
Celite 281	25	1.30
W.G. Mica	25	1.07
Water	335	40.02
KTPP	2	0.11
Strodex PK-90	8	0.83
Igepal CO-630	5	0.57
Colloids 581-B	2	0.20
Propylene Glycol	25	2.68
Cellosize QP-4400	4	0.48
Ammonium Hydroxide (28%)	1	0.12
Aroplaz 1271 Alkyd	64	7.66
Advacar Lead (24%)	2	0.21
Advacar Cobalt (6%)	1	0.13
Elvace PB-3-1952	284	31.20
lb/gal		12.21
Nonvolatile (%)		58.43
PVC (%)		40.50

Self-Priming Exterior Latex Light Tint Base

Water	308
KTPP	2
Daxad 30	9
Igepal CO 630	3
Advacide 60	1.5
Defoamer	2
Celite 281	35
Nytal 300	100
Titanox RA-NC	225
EP 303 Lead Silicate	133
T&W 345-100 Super Alkyd	62
Lead Drier (24%)	1
Cobalt Drier (6%)	.4
Ethylene Glycol	25
Carbitol Acetate	9
Natrosol 250 MR	4

Parco 37-C-55	310
Yield (gal)	100.9
PVC (%)	40.0
Solids (%)	58.8
Visc. (KU)	88–94
lb/gal	12.3

Exterior Tint Base

(Acrylic)

	lb	gal
Water	220.0	26.47
Latekoll D (8%)	50.0	5.92
Daxad 30	8.0	.77
Igepal CO-630	3.0	.38
PMA-30	5.0	.52
Titanox RANC	200.0	5.72
Minex #7	100.0	4.59
Nytal 300	65.0	2.74

Add in order listed and mix on a high speed disperser for 10 min, then let down as follows:

Texanol	21.0	2.65
Ethylene Glycol	37.0	4.00
Acronal 290 D	420.0	48.28
Deefo 97-2	2.0	.23
PVC (%)		34.9
Nonvolatile (%)		52.5
Nonvolatile by vol. (%)		37.4
Wt./gal (lb)		11.3
Visc. (KU)		80±5

Exterior Light Tint Base-Alkyd Modified

FORMULA NO. 1

(Vinyl Acrylic)

	lb	gal
Pigment Grind:		
Water	225.0	27.0
Cellosize Hydroxyethyl Cellulose		
QP-4400H	5.0	0.4
Dowicil 75	1.5	—
Tamol 850	3.0	0.3
Victawet 35B	7.0	0.7
Igepal CTA-639	3.0	0.3
Ethylene Glycol	25.0	2.7
Butyl Carbitol	12.0	1.4
Nopco NXZ	2.0	0.2
Nopococide N-96	4.0	0.4
Horse Head XX-503	75.0	1.6
Ti-Pure R-966	175.0	5.0
Mica, 325 mesh W.G.	25.0	1.1
Snowflake	150.0	6.6
Let Down:		
Igepal CTA-639	5.0	0.6
Ucar Latex 365	312.0	34.4
Aroplaz 1271 ⎫	43.0	5.2
Zirconium Drier (6%) ⎬ Premix	1.5	—
Cobalt Drier (6%) ⎭	1.5	—
Nopco NXZ	2.0	0.3
Water/QP-4400H 2-1/2% Sol'n.	98.5	11.8

PVC (%)	38
Total Solids (%)	56.4
Alkyd Modification (%)	20
Visc. (KU)	80–90
Wt./gal	11.7–11.8

Note: Protect Solutions of Cellosize HEC with preservative.

No. 2

(Vinyl – Acrylic)

Water	120
Tamol 731	15
Triton CF-10	2.5
Dapro 881-S	1
Metasol 57	1.8
Titanox CL-NC	175
Imsil A-25	250

Mix on a high speed mill and add:

Texanol	7
Ethylene Glycol	25
Natrosol 250 MR	3.5
T&W 345-100 Super Alkyd	30
Lead (24%)	1
Manganese (6%)	0.5
Dapro 881-S	1
ParCryl 400	410
Propylene Glycol	35
Ammonium Hydroxide	2
Triton N-57	4
Triton X-207	4

Yield (gal)	100
PVC (%)	40
Solids (%)	54
Visc. (KU)	74–78
pH	9.5

Exterior Deep-Tone Tint Base

(Vinyl – Acrylic)

	lb	gal
Charge into mixer under agitation:		
Water	60.0	7.2
Cosan PMA-30	2.0	0.2
Ethylene Glycol	25.0	2.6
Igepal CO-630	3.0	0.3

	lb	gal
Tamol 731	4.0	0.4
Potassium Tripolyphosphate	1.0	0.1
Colloid 677	1.0	0.1
Texanol	8.0	1.0
Natrosol 250 HR (1½% Sol'n.)	70.0	8.4
Titanox RANC	75.0	2.2
Snowflake	100.0	4.4
ASP-400	75.0	3.5
Celite 281	50.0	2.6

Disperse in high-speed mill, then add slowly with agitation:

	lb	gal
Resyn 2345	309.0	34.0
Igepal CO-630	5.0	0.6
Advacar 6% Cobalt Drier	1.5	0.2
Advacar 24% Lead Drier Premix,	3.0	0.3
Extra Long Oil Alkyd then add		
(100% Solids)	55.0	6.6
PMO-30	3.0	0.3
Natrosol 250 HR (1½% Sol'n.)	180.0	21.4
Colloid 677	1.0	0.1
Water	30.0	3.6

PVC (%)	35
CPVC (%)	46
Wt./gal (lb)	10.6
Visc. (KU)	75–80
85° Angular Sheen	1
Total Solids (%)	51.2
Modifier (%)	25

Note: Protect thickener solution with 0.08% **PMA-30** if stored prior to use.

Exterior Acrylic Latex Tint Base

	lb	gal
Dispersion:		
Water	206.5	24.79

	lb	gal
Super Ad-It	8.0	1.00
Nopco NXZ	2.0	0.30
Ethylene Glycol	27.8	3.00
Ammonium Hydroxide (28%)	1.0	0.10
Methocel 65 HG-DG (4000 cps)	3.0	0.25
Potassium Tripolyphosphate (20% Sol'n.)	5.0	0.50
Tamol 731	4.0	0.44
Titanox CL-NC	200.0	5.87
Natural Whiting	180.0	8.03

Let Down

	lb	gal
Igepal CO-630	5.0	0.57
Texanol	7.9	1.00
Rhoplex AC-35 (46% Solids)	423.0	48.00

Premix and add:

	lb	gal
Castung 235	47.1	6.00
Zirco Catalyst (6% Zr)	0.6	0.08
Cobalt Naphthenate (6% Co)	0.6	0.07

Pigment (%)	33.8
PVC (%)	35.0
Vehicle (%)	66.2
Nonvolatile (%)	33.2
Total Solids (%)	55.6

Tint Base Exterior Gloss Trim

(Vinyl – Acrylic)

Propylene Glycol	70
Tamol 731	16
Triton CF-10	1
Dapro 881-S Defoamer	1
Water	30
Du Pont R-900	200
Barytes	35

Mix on high speed mill and add

Texanol	4
ParCryl 400	515
Dapro 881-S	1
Ammonium Hydroxide	2
Propylene Glycol	50
Water	100
Acrysol G-110 (as needed)	10
Triton N-57	4
Triton X-207	4
Yield (gal.)	100
PVC (%)	20
Solids (%)	47
Visc. (KU)	68–72
pH	9.5

Exterior Acrylic Tint Base

FORMULA NO. 1

	lb	gal
Premix:		
Water	100.0	12.00
Tergitol NPX Dispersing Agent	3.0	0.34
AMP 2-Amino-2-Methyl-1-Propanol	6.0	0.78
Oleic Acid	2.0	0.27
Increase speed and add slowly:		
Chemacoil TA-100	65.0	8.33
Disperse pigments:		
Titanox RA-NC Titanium Dioxide	200.0	5.72
Mica (235-mesh water-ground)	25.0	1.16
Duramite Calcium Carbonate	150.0	6.65
Decrease speed and add slowly:		
Nopco NDW Antifoam	4.0	0.40
Ethylene Glycol	18.6	2.00
Nuodex PMA-18 Fungicide	8.0	0.80
Cellosize QP-15000 (2%) Thickener	156.0	18.55
Rhoplex AC-34 Acrylic Emulsion (46%)	330.0	37.08
Super-Cobalt Drier	1.0	0.13

Visc. (KU)	68
pH	9.0
PVC (%)	35
Total Solids (% by Wt.)	54.9
(% by vol.)	40.5

Chemacoil TA-100 = 30% of binder solids

No. 2

(Deep-Tone for Universal Colors)

Water	162
Ethylene Glycol	35
Oleic Acid	7
KTPP	3
Natrosol 250 MR	3
Titanox CL-NC	35
Gold Bond R	300
Advacide 60	2
ParCryl 300	487
Acrysol G-110	6
Water	18
Ammonium Hydroxide (28%)	6
Yield (gal)	100.3
PVC (%)	40.0
Solids (%)	53.8
Visc. (KU)	81–87
pH	9.0–9.6

Exterior Poly Vinyl Acetate Tint Base

FORMULA NO. 1

	lb	gal
Dispersion:		
Water	150.0	18.03
Carbitol	30.0	3.51
Daxad 30	5.0	0.52
Igepal CA360	5.0	0.57

	lb	gal
Ti-Pure Titanium Dioxide R-960	175.0	5.32
Nytal 300	116.0	4.88
Mica 325 (mesh waterground)	30.0	1.27
Celite 281	50.0	2.60
Colloid 581-B	1.0	0.15

Reduction:

	lb	gal
Cellosize WP-4400 (2% sol'n.)	118.0	14.10
Colloid 581-B	1.0	0.15
Witco 912	4.0	0.48
Elvacet 1454	432.0	48.00
2-Amino-2-Methyl-1-Propanol	4.0	0.51
Super-Ad-It	10.0	1.23

PVC (%)	36.4
Visc. (KU) — overnight	70–75

Disperse in high speed disperser.

No. 2

	lb	gal
Premix:		
Water	150.0	18.00
AMP 2-Amino-2-Methyl-1-Propanol	3.0	0.39
Oleic Acid	2.0	0.54
Potassium Tripolyphosphate	1.0	0.05

Increase speed and add slowly:

	lb	gal
Chemacoil TA-100	150.0	19.23

Disperse pigments:

	lb	gal
Ti-Pure R-901 Titanium Dioxide	125.0	3.57
Mica (325 mesh water-ground)	25.0	1.06
Celite 281 Diatomaceous Silica	50.0	2.60
Duramite Calcium Carbonate	250.0	11.08

	lb		gal
Decrease speed and add slowly:			
Tributyl Phosphate/Pine Oil			
(3:1 by Wt.)	3.0		0.36
Ethylene Glycol	18.6		2.00
Nuodex PMA-18 Fungicide	3.0		0.30
Everflex BG Vinyl Acetate Copolymer			
Emulsion (52%)	266.0		29.89
Super-Cobalt Drier	1.5		0.19
Cellosize QP-15000 (2%) Thickener	100.0		11.91
Visc. (KU)		69	
pH		8.4	
PVC (%)		35	
Total Solids (% by Wt.)		65.0	
(% by vol.)		53.0	
Chemacoil TA-100 = 52% of binder solids			

No. 3

	lb	gal
Premix:		
Water	141.0	17.00
Makon 10 Surfactant	7.2	0.90
Clearate WD Lecithin	8.0	1.00
Triton X-100 Surfactant	4.5	0.56
Increase speed and add slowly:		
Chemacoil TA-100	150.0	19.23
Disperse pigments:		
Titanox RA-NC Titanium Dioxide	125.0	3.58
Azo ZZZ33 Zinc Oxide	50.0	1.07
Celite 281 Diatomaceous Silica	50.0	2.60
Mica (325-mesh water-ground)	25.0	1.16
Duramite Calcium Carbonate	225.0	9.96
Slow mixing speed and add slowly:		
Colloid 581-B Defoaming Agent	2.0	0.25

	lb		gal
Ethylene Glycol	18.6		2.00
Nuodex PMA-18 Fungicide	3.0		0.30
Cellosize QP-15000 (2%) Thickener	92.0		11.00
Everflex BG Vinyl Acetate Copolymer Emulsion (52%)	267.0		31.00
Super-Cobalt Drier	1.5		0.13
Ammonia (28%)	4.0		0.50
Visc. (KU) — overnight		82	
3 months		82	
PVC (%)		35	

Chemacoil TA-100 = 52% of binder solids

Exterior Latex Paint Tint Base

(Poly Vinyl Chloride)

	lb	gal
Water	140.0	16.8
Triton X-100	3.0	.3
Ethylene Glycol	20.0	2.1
1-Amino-2-Methyl-1-Propanol	4.5	.5
Tamol 731	9.0	1.0
Hercules 357	6.0	.8
Titanox RANC	180.0	5.1
Celite 281	50.0	1.6
211 Mica	30.0	1.3
Duramite	100.0	4.4
Water	25.0	3.0
Potassium Tripolyphosphate	2.0	—
Cellosize WP-4400 Sol'n. (2½% Sol'n.)	100.0	12.0
Castung 235 Intermediate (BFG-12279)*	153.7	19.1
Triton X-100*	5.0	.5
Geon 450 X20*	260.0	27.6
Water or **WP-4400** Sol'n.	17.5	2.1

Castung 235 Intermediate (BF-12279)*

Castung 235	433.4	55.3
Cobalt Naphthenate (6%)	2.1	.3
Lead Naphthenate (24%)	7.3	.8

	lb	gal
Ammonium Hydroxide (18%)	15.7	2.1
Water	326.4	39.2
Igepal CO-630	20.9	2.3

PVC (%)	34.92
Nonvolatile (%)	52.01
Nonvolative vol. (%)	36.14
Wt./gal (lb)	11.35
Visc. (KU)	75–80
pH	9.1
Freeze Thaw Cycles	7
Oven Stability (140 F)	1 month
˙Color Acceptance	Excellent

This formula has been tested with five different "universal" colorant systems and found to accept all of them with ease. Tints of these systems were tested for pH, viscosity and color drift, and found to be stable in each.

---

*Premix before incorporating in pigment slurry.

Chapter III

MARINE PAINTS

Marine paints must withstand constant exposure and, for a large percentage of them, immersion in salt water. Many contain antifouling ingredients that must be scrutinized in light of constantly increasing environmental sanctions. Coatings are usually within the 7–12 mil thickness range after from 4–7 applications of high-build paint. Application varies from a paint brush to underwater airless spray systems.

Marine Anticorrosive Primer High Flash Point Solvent

(Vinyl)

Gantrez VC	17.62
Aluminum Stearate	0.11
Red Lead	27.53
Xylene	54.74
PVC (%)	17.44
Theoretical 1-mil coverage (ft^2/gal)	355.27
Nonvolatiles by Wt. (%)	45.26
Nonvolatiles by vol. (%)	22.15
Visc. (KU)	84
Wt. per gal	10.31
Specific gravity	1.24
Flash Point (°F)	81

Marine Anti-Fouling Ship-Bottom Paint with High Flash Point

(Vinyl)

Gantrez VC	2.84
Resin WW	11.09
Bentone 38	0.31
Cuprous Oxide	74.30
Xylene	11.46
PCV (%)	49.20
Theoretical 1-mil coverage (ft^2/gal)	1048.13
Nonvolatiles by Wt. (%)	88.54
Nonvolatiles by vol. (%)	65.34
Visc. (KU)	95
Wt. per gal	21.88
Specific gravity	2.63
Flash point (°F)	81

Marine Vinyl/Alkyd Enamel (Black)

Gantrez VC	10.56
Rezyl 869	23.92
Lampblack	10.56
Driers Mix[1]	0.06
Xylene	54.90
PVC (%)	16.16
Theoretical 1-mil coverage (ft^2/gal)	590.41
Nonvolatiles by Wt. (%)	45.08
Nonvolatiles by vol. (%)	36.81
Visc. (KU)	70
Wt. per gal	8.35
Specific gravity	1.00

[1] 0.4 lead/0.04 cobalt/0.04 manganese % alkyd solids

Grey Marine Vinyl/Alkyd Enamel

Gantrez VC	9.48
Rezyl 869	22.14
Titanium Dioxide	3.79

Zinc Oxide XX-600	13.91
Lampblack	0.15
Talc	10.12
Driers Mix[1]	0.05
Xylene	40.30
PVC (%)	19.96
Theoretical 1-mil coverage (ft^2/gal)	693.31
Nonvolatiles by Wt. (%)	59.62
Nonvolatiles by vol. (%)	43.22
Visc. (KU)	71
Wt. per gal	10.2
Specific gravity	1.22

---

[1] 0.4 lead/0.04 cobalt/0.04 manganese % alkyd solids

Zinc Dust Primer

(Epoxy)

Base Vehicle:

Epon 1001	4.46
Thixatrol ST	0.36
Beetle 216-8	0.40
Mixed Solvents	8.00

Curing Agent:

Versamid 115	2.03
Mixed Solvents	3.33

Pigment:

Zinc Dust	81.42

Procedure

Dissolve the **Epon** resin in the mixed solvents to produce a solution at 60% solids. Add the **Beetle** resin solution and then disperse the **Thixatrol ST**, so that a temperature of about 120°F is obtained. The remaining solvent is then added to produce the vehicle base component.

The curing agent solution is best prepared by warming the **Versamid 115** and pouring it (when fluid) into the mixed solvents.

The paint is prepared by mixing the vehicle base component with the curing agent and then adding the zinc dust with agitation.

Underwater Coating

(Epoxy)

	lb
Araldite 6010	314.5
Silica Sand	146.7
Titanium Dioxide	26.2
Amorphous Silica	75.5
CIBA Polyamide 825	283.0
Accelerator 064	15.7
Silica Sand	167.7
Amorphous Silica	52.4
Carbon Black	0.5

Compositions of Rosin-Vinyl Antifouling Paints
(Nonvolatile Portion Only)

FORMULA	No. 1	No. 2
Toxicant:		
Cupric Oxide	207.5	104
Copper Pigment	207.5	104
Organolead	—	73.7
Binder:	(100)	(100)
Vinyl	16.7	16.7
Rosin	68.5	68.5
[1] TCP	14.8	14.8
Filler or Pigment:		
Black	25.8	25.8
Barytes	—	120.0

[1] Tritolyl Phosphate

Compositions of Rosin-Vinyl Antifouling Paints (cont'd)
(Nonvolatile Portion Only)

	No. 3	No. 4
Toxicant:		
Cupric Oxide	421	450
Binder:	(100)	(100)
Vinyl	42.1	17.2
Rosin	42.1	67.2
TCP*	15.8	15.6
Rosin-Vinyl Ratio:		
By Wt.	1/1	3.9/1

---

*Tritolyl Phosphate

Rosin Antifouling Coatings

	FORMULA No. 1	No. 2
Rosin	50.0	50.0
Paraffin Wax	3.0	3.0
Microcrystalline Wax	6.0	6.0
Oleic Acid	6.0	6.0
Oil Soluble Nigrosine	2.5	2.5
DDT	1.3	0
Fouling Resistance (%)		
2 months	90.0	0
8 months	80.0	0
12 months	0	0

It is well established that DDT is a specific contact nerve toxin for arthropods and acts by being absorbed through their joints.

Conclusions:

1. The presently available, or "1st Generation" antifouling coatings are based on toxins, such as mercuric oxide, cuprous oxide, tributyltin oxide, tributyltin fluoride, 10, 10'-oxybisphenoxyarsine and triphenyl lead acetate. These toxins slowly dissolve and leach out of the matrix creating a toxic environment near the surface of the coating.

Chlorinated Rubber Marine High-Build Coatings with Inhibiting Pigment for Airless Spray

	FORMULA No. 1 (Red Lead)	No. 2 (Aluminum/ Basic Lead Sulfate)	No. 3 (Aluminum)
Chlorinated Rubber (10 cps)	12.1	12.3	12.0
Chlorinated Paraffin 42	4.0	4.1	4.0
Chlorinated Paraffin 70	7.1	7.2	7.0
Red Lead	18.2	—	—
Red Iron Oxide	18.2	6.7	0.3
Basic Lead Sulfate	—	13.3	—
Barytes	—	6.7	18.0
Nonleafing Aluminum Paste (65%)	—	9.2	12.0
Titanium Dioxide	—	—	6.0
Modified Hydrogenated Castor Oil	1.6	1.6	1.6
Xylene	31.0	31.1	30.5
Mixed Trimethyl Benzenes	7.8	7.8	7.6

Of the high build primers studied, red lead is apparently the worst, with basic lead sulfate/aluminum marginally better than that containing only aluminum.

	No. 4 (Red Lead)	No. 5 (Aluminum/ Basic Lead Sulfate)	No. 6 (Aluminum)
Alloprene R10	12.1	12.3	12.0
Cereclor 42	4.0	4.1	4.0
Cereclor 70	7.1	7.2	7.0
NS Red Lead	18.2	—	—
Spanish Red Oxide	18.2	6.7	—
Basic Lead Sulfate	—	13.3	—
Barytes	—	6.7	18.0
Nonleafing Aluminum Paste (65%)	—	9.2	12.0

	No. 4	No. 5	No. 6
		(Aluminum/ Basic Lead	
	(Red Lead)	Sulfate)	(Aluminum)
Tioxide R-CR	–	–	6.0
Deanox U90	–	–	1.0
Magnesium Oxide	–	–	0.3
Thixatrol ST	1.6	1.6	1.6
Xylene	31.0	31.1	30.5
Aromasol H	7.8	7.8	7.6
PVC (%)	33.5	31.0	34.2
Wt./gal (lb)	14.9	13.8	13.4

Metallic Lead Primer

(Chlorinated Rubber)

Chlorinated Rubber (20 cps)	11.0
Chlorinated paraffin 42	4.4
Micronized Talc	4.1
Barytes	19.6
Metallic lead paste (91% in chlorinated paraffin 42)	32.0
Modified hydrogenated castor oil	0.36
Solvent blend of mixed Trimethyl Benzenes (b.p. 168–205°C) and Mineral Spirits in 3:1 ratio	28.4

Chlorinated Rubber Antifouling Coatings

	Formula No. 1	No. 2
Alloprene 20 cps	7.9	6.6
Wood Rosin WW	7.3	4.8
Tricresyl Phosphate	3.0	2.2
Thixatrol ST	0.6	0.2
Cuprous Oxide	40.1	56.9
Zinc Oxide	–	3.5
Mercuric Oxide	2.5	–
Red Iron Oxide	2.9	–
Magnesium Oxide	2.0	–
Solvesso 100	25.3	20.6
Short range spirits	8.4	5.2

	No. 1	No. 2
WPG lb	13.2	16.2
PVC (%)	38.0	50.0
Visc. (KU)	73	93
NV Wt. (%)	66.3	74.2
NV vol (%)	37.0	41.3

Formula No. 1 is a standard antifouling coating, while Formula No. 2 represents a recently developed high cuprous oxide composition offering greater durability to match the needs of the thick coating systems.

These antifouling coatings are suitable for brush or air assisted spray application. They have not been tested for airless spray application.

No. 3

Alloprene 20	10.0
Wood Rosin	10.0
Cereclor 42	5.0
Thixatrol ST	0.3
Aromasol H*	29.2
Mineral Spirits	9.5
Spanish Red Oxide	5.0
China Clay	16.0
Mercuric Oxide	1.0
Cuprous Oxide	14.0

The general view is that for continuous leaching of active constituent 60% of the binder should be rosin. The cuprous oxide content (or mercury content) depends on the requirements and for waters in temperate climes, the following formulation has been found satisfactory:

For tropical waters the cuprous oxide must be increased very considerably, as also the mercuric oxide — or other mercury compound. Use of fungicides is also recommended for tropical waters.

---

*A blend of Solvesso 100 and Solvesso 150 may be substituted for the Aromasol H requirement.

Chlorinated Rubber Antifouling Marine Paint System

(Metallic Lead Primer)

Alloprene 20	15.2
Cereclor 42	6.7
Metallic Lead Paste (90% in **Cereclor**)	19.0
Micronized Barytes	9.5
Micronized Red Iron Oxide (synthetic)	4.8
C-3000 Mica	4.8
Thixatrol ST	0.5
Epoxidized Soya Oil	0.9
High-Flash Aromatic Solvent	29.0
Mineral Spirits	9.6

On top of the primer coat, two applications by airless spray of an **Alloprene** thick coating are recommended giving an overall protective coating thickness of about 7 to 12 mils. It is further suggested that contrasting colors be used in each coat to minimize "thin spots" in painting. Thick coating formulations for airless spraying are given below.

(Thick Coatings)

	Red	Gray
Alloprene 10	12.5	12.5
Cereclor 42	4.1	4.1
Cereclor 70	7.4	7.4
Rutile Titanium Dioxide	–	12.5
Red Iron Oxide (natural)	10.0	–
Carbon Black	–	0.2
Barytes	15.0	12.5
Thixatrol ST	1.6	1.6
Xylene	39.4	39.2
2-Ethoxy Ethyl Acetate	10.0	10.0

A thixotropic antifouling paint based on **Alloprene** with good keying properties to the thick coatings is shown below.

(Antifouling Topcoat)

Alloprene 20	8.2
American Wood Rosin	8.2

Cereclor 42	2.0
Red Iron Oxide	4.0
Cuprous Oxide	26.0
Mercuric Oxide	1.0
Magnesium Oxide	4.0
China Clay	8.0
Thixatrol ST	0.6
High Flash Aromatic Solvent	29.0
Mineral Spirits	9.0

The presence of magnesium oxide in the antifouling paint improves its can stability. The specific formulation given above has been stored satisfactorily in cans for 120 days at 50°C without gelation.

Marine Coating for Cold, Damp Weather Application

(Polyurethane – Epoxy)

		lb	gal
A	Black Iron Oxide (#724 Black Iron Oxide)	125	3.0
	Carbon Black (Black Pearls #81)	50	3.2
	Phthalocyanine Blue	9	0.7
	Magnesium Silicate (**Nytal 300**)	175	7.4
	Celite 281	75	3.9
	Epoxy Resin (**Epon 1004**)	59	6.0
	Castor 1066	72	9.0
	Ethyl Acetate	121	16.2
	Toluene	117	16.2
	2-Nitropropane	137	16.2
B	**Spenkel P49-75S**	176	28.0

Visc. (KU)	61
PVC (%)	39.0
Vehicle Solids (%)	38.6
Total Solids (%)	60.5
Wt./gal	11.2
NCO to OH ratio	0.95–1.0

Procedure:

For warmer environment, slower evaporating solvents should be used. B is added to A and mixed thoroughly for 15–30 min. Let stand for 30 min before use, at 35–40°F. Solvents as indicated in A can be used to reduce for spraying, if necessary.

Marine Primer for Cold, Damp Weather Application

(Urethane – Epoxy)

		lb	gal
A	Basic lead silico chromate	448	13.2
	Black iron oxide	42	1.0
	Magnesium silicate	55	2.3
	Thixatrol ST	6	0.7
	Mica	71	3.0
	Epoxy resin **Epon 1004**	63	6.5
	Polycin 53	78	9.7
	Ethyl acetate (urethane grade)	76	10.1
	Toluene	73	10.1
	2-Nitropropane	86	10.1
B	**Vorite 144**	188	19.2
	Ethyl acetate (urethane grade)	35	4.7
	Toluene	34	4.7
	2-Nitropropane	40	4.7

PVC (%)	40 ± 2
Visc. @ 35°F (KU)	72–80
Grind (minimum)	5
Density (lb/gal)	12.7–13.3
Nonvolatile (%)	69–73
Flash Point (min) °F	75
Gloss, 60° (max)	10
Pot Life @ 35°F, sealed (hr)	30
Dry (35°F, 75% RH):	
Set-to-Touch (min)	40
Dry Hard (hr)	8
Full Cure (hr)	24

Application:

This **Vorite 144–epoxy–Polycin 53** system is a development based on recent studies by the U.S. Naval Research Laboratory. N.R.L. Report 6308 describes a urethane–epoxy coating for use as a submarine coating for application in cold, damp weather. Navy performance requirements call for at least six months durable performance either as a topside or submerged coating (exposure to water at high flow rates).

Thixatrol ST is recommended in the primer formulation in order to eliminate the pigment settling and impart excellent sag control, thixotropic body, significant viscosity increase and improved application properties.

Tri-Amino Marine Enamel

FORMULA No. 1

(Black)

	lb	gal
Carbon Black	20.0	1.36
Aluminum Stearate	2.0	0.24
Bentone 38 Gel (10% in mineral spirits)	20.0	2.85
Chemacoil TA-303 Resin (80%)	412.0	54.00
Carbamac 39 Oil-Modified Urethane (50%)	219.8	29.50
Mineral Spirits	60.1	9.25
Cobalt Drier (6%)	2.0	0.25
Lead Drier (24%)	6.0	0.62
Zinc Drier (8%)	7.5	0.80
Antiskinning Agent	2.0	0.26
Maxvar 2503 Additive (55%)	25.0	3.25

Visc. — Fresh (KU)	75
PVC (%)	2.5
Vehicle Nonvolatile (% by Wt.)	60.6
Wt. per gal	7.58

No. 2

(White)

	lb	gal
Titanox CL-NC Titanium Dioxide	250.0	7.50
Aluminum Stearate	2.0	0.24
Bentone 38 Gel (10% in mineral spirits)	20.0	2.85
Chemacoil TA-303 Resin (80%)	300.0	39.50
Carbamac 39 Polyurethane (50%)	160.0	21.50
Mineral Spirits	188.0	29.00
Cobalt Drier (6%)	2.0	0.25
Zirconium Drier (6%)	5.0	0.62
Antiskinning Agent	2.0	0.26
Maxvar 2503 Additive (55%)	25.0	3.25

Visc. — Fresh (KU)	82.1
PVC (%)	15.9
Vehicle Nonvolatile (% by Wt.)	49.4
Wt. per gal	9.09

No. 3

(Green)

	lb	gal
A-4558 Monarchrome Green L Pigment	150.0	6.30
Aluminum Stearate	2.0	0.24
Bentone 38 Gel (10% in mineral spirits)	20.0	2.85
Chemacoil TA-303 Resin (80%)	393.0	52.50
Carbamac 39 Oil-Modified Urethane (50%)	210.5	28.25
Mineral Spirits	65.0	10.00
Cobalt Drier (6%)	2.0	0.25
Zirconium Drier (6%)	5.0	0.62
Antiskinning Agent	2.0	0.26
Mavar 2503 Additive (55%)	25.0	3.25

Visc. — Fresh (KU)	81.0

PVC (%)	10.9
Vehicle Nonvolatile (%)	59.4
Wt. per gal/lb.	8.45

These formulations are three variations of an outstanding marine enamel based on **Chemacoil TA-303** plus **Carbamac 39**, an oil-modified urethane resin. They have been evaluated on boats in both fresh and salt water, and show excellent gloss retention. In all instances the finish has lasted more than one season without necessity for repainting. Boat owners also appreciate the ease of cleanup when boats are withdrawn from the water.

Tri-Amino Marine Spar Varnish

	lb	gal
Heat to $130°F$:		
Chemacoil TA-303 Resin (80%)	442.0	58.00
Add and stir to dissolve:		
Uvinul 400 uv Absorber	17.5	1.48
Add remaining ingredients:		
Mineral Spirits	260.0	40.00
Cobalt Drier (6%)	3.0	0.37
Zirconium Drier (6%)	6.0	0.83
Antiskinning Agent	1.0	0.14
Visc.		D Gardner
Color		9 Gardner
Total Solids (% by Wt.)		51
Wt. per gal/lb		7.23

This varnish, along with two other widely accepted marine spar varnishes, were exposed on redwood and yellow pine panels at 45° south in Florida and Terre Haute, Indiana, and on a floating testrack in Florida. The results of observation of gloss and checking (10 = no checking; 0 = badly checked) are given below.

Exposure Time	This Formula		Marine Spar A		Marine Spar B	
	Gloss	Check	Gloss	Check	Gloss	Check
3 months	74	10	71	10	74	10
6 months	74	10	55	2	74	6
9 months	63	5	38	0	23	0

Chapter IV

METAL PAINTS

Aside from the specific substrate preparations indicated for each metal or alloy, the most important distinction among metal paints is their drying and curing schedules. Two main groups are air-dry solvent or water systems and bake preparations. The latter group is formulated for and serves industrial needs.

Ready-Mix Aluminum Paint

(Linseed Oil)

	lb
Aluminus Paste (65.5% N.V.)	200.8
PC-3470 (55.6% N.V.)	618.8
Cobalt Drier (6%)	1.9

Galvanized Metal Primer

Formula No. 1

(Alkyd)

	lb	gal
Titanox C-50	200	6.92
Zinc Oxide (**Kadox 515**)	25	0.53
Whiting (5-25)	175	7.79
Barytes	50	1.35
Varcopol 428-50X	335	40.80

	lb	gal
Xylene } Preblend Let Down	150	20.70
VM&P Naphtha } solvents – then add	125	20.30
Cobalt Naphthenate (6%)	1.0	0.13
ASA	0.5	–
PVC (%)	48.6	
Vehicle Nonvolatile (%)	27.5	
Total Nonvolatile (%)	58.0	
Wt. per gal (lb)	10.8	
Visc. (KU)	67–77	

No. 2

(Alkyd)

	lb	gal
Basic Silicate White Lead (**Oncor 45X**)	250	7.50
Zinc Chromate	50	1.73
Desertalc 55	150	6.50
Varkyd 303-50 VM	335	44.40
VM&P Naphtha	275	44.40
Cobalt Naphthenate (6%)	1.0	0.13
Lead Naphthenate (24%)	2.5	0.26
ASA	0.5	–
PVC (%)	46.2	
Vehicle N.V. (%)	27.5	
Total N.V. (%)	58.0	
lb/gal	10.2	
Visc. (KU)	67–77	

Zinc Chromate Primer

(Alkyd)

Pure Zinc Chromate	702.0
Magnesium Silicate	120.0
Aroplaz 1365-X-60	880.0
Maleic Anhydride	2.5
Bakelite CKU-5062 (50% Solids)	225.0

Xylene	548.0
Post-4*	9.0
Lead Naphthenate (24%)	10.0
Cobalt Naphthenate (6%)	2.5
Antiskinning Agent	2.5

Specification of Suspension Properties:

In accordance with the specification, each of the two primers (with and without **Post-4** were thinned for spraying (one volume primer to a two volume of toluene).

After a 24 hr aging period, the control paint exhibited hard settling whereas the paint with **Post-4** had no settling at all.

---

*Post-4 added in the grind.

Lower Cost Utility Red Oxide Primer

(Alkyd)

	lb	gal
Varkyd 363-50X	370	45.12
Post 4	2	.20
TDO	4	.50
Lead Napthenate (24% active)	3	.10
1503-RO	220	5.95
Zinc Yellow	11	.35
Bentone 38	6	.30
CP 14-35	110	4.87
Vicron 15-15	75	3.31
#9 Clay	75	3.47
Toluene	85	11.74
Standard 250	150	23.83
Cobalt **Nuxtra**	1.25	.15
Manganese **Nuxtra**	1.33	.16
ASA	1	.13

PVC (%)	47.0
Vehicle Nonvolatile by Wt. (%)	30.5
Total N.V. by Wt. (%)	61.7
Wt. per gal (lb)	11.1
Visc. (KU)	70

Corrosion Resistant Top Coat

FORMULA NO. 1

(Alkyd – No Color)

	lb/100 gal
#316 Micaceous Iron Oxide	630
#38 **Bentone** Mastergel (15%)	79
P-296-60 Soya Alkyd	160
Mineral Spirits	108

Disperse in a Hockmeyer.

Add:

P-296-60 Alkyd	160
Mineral Spirits	107
Troykyd (6% Cobalt Naphthenate)	2.2
Troykyd (5% Calcium Naphthenate)	0.7
Troykyd (24% Lead Naphthenate)	5.2
Troykyd (Antiskin)	2.3

Solids by vol.	41.76
PVC (%)	35
Wt./gal	12.5
Visc. (KU)	68
Pigment-Vehicle Ratio	67.5/32.5
Volatile	31.9
Hiding Power	334 ft^2/gal
	40 ft^2/lb

4-Mil (dry) film coat over 1-Mil (dry) anticorrosive primer suggested.

NO. 2

(Alkyd – Green)

	lb/100 gal
#316 Micaceous Iron Oxide	505
529 Green Chromium Oxide	90
P-296-60 Soya Alkyd	190
Aluminum Paste (65% NV)	50
Cabosil M-5	6 3/4

	lb/100 gal
#38 **Bentone** Mastergel (15%)	75
Mineral Spirits	22

Disperse in Hockmeyer.

Add:

P-296-60 Soya Alkyd	230
Mineral Spirits	90
Troykyd (6% Cobalt Naphthenate)	2.8
Troykyd (5% Calcium Naphthenate)	1.8
Troykyd (24% Lead Naphthenate)	8.0
Troykyd (Antiskin)	4.0
#316 Micaceous Iron Oxide lb/gal	5.05
Solids by vol. (%)	49.4
PVC (%)	34.8
Wt. per gal lb	12.76
Visc. (KU)	83
Pigment Vehicle Ratio (%)	51–49
Volatile (%)	28
Hiding Power	1450 ft^2/gal
	227 ft^2//lb

4-Mil (dry) film coat over 1-Mil (dry) anticorrosive primer suggested.

Deep Yellow Aersol Enamel

(Alkyd)

	lb
Base:	
Medium Chrome Yellow	200
Molybdate Orange	20
X-2280 IAF Compound	2.0
Post 4	8.7
Syntex 3638	350.0
Xylene	57.0

Pebble mill to 9–10 paint club scale reading, then add:

			lb
Syntex 3638			350
Zirco Drier (6%)			5.9
Cobalt Naphthenate (6%)			3.0
Exkin #2			2.0
Antimar Agent			2.0

Reduction:

			lb
Base			556.0
Toluene			322.0

Reduction Visc.: 11–13 s (Ford Cup #)

Can Loading:

Reduced Paint	53.4		50.7
Propellant P	46.6		49.3
Visc. (KU)		106	
Wt./gal		10.0	
PVC (%)		10.7	
Vehicle Solids (%)		46.0	
Total Solids (%)		58.4	

Gray Automotive Enamel

(Alkyd)

	lb
Roller Mill Grind:	
Titanium Dioxide	63.0
Lamp Black	1.3
Duraplex A-29 (60% solids)	107.0
Mix with:	
Duraplex A-29 (60% solids)	349.0
Uformite MM-46 (60% solids)	151.3
Xylene	185.5
Wt. per gal	8.6 lb
Total Solids	50.0%

	lb
Pigment:	15.0%
Titanium Dioxide	98.0%
Lamp Black	2.0%
Vehicle:	85.0%
Duraplex A-29	75.0%
Uformite MM-46	25.0%

At 50% solids, the above formulation has a viscosity of 75–80 s in a No. 4 Ford Cup. It should be reduced with xylene to about 40% solids for spray. Suggested curing schedule is 30 min at 250 F, or 15 min at 300 F.

White Enamel

(Alkyd)

	lb	gal
Roller mill grind:		
Titanium Dioxide	355	10.9
Duraplex ND-78 (60% solids)	236	27.8
Mix with:		
Duraplex ND-78 (60% solids)	177	20.7
Uformite MX-61 (60% solids)	177	20.4
Xylene	146	20.2
Wt./gal (lb)		10.9
Total Solids (%)		65
Pigment		50%
Binder		50%
Uformite MX-61		30%
Duraplex ND-78		70%

The enamel as made at 65% solids is at storage viscosity. It will spray at 58% solids. The suggested baking schedule is 30 min at 300 F.

This enamel, when applied in a single coat over bonderized steel and baked for 30 min at 300 F., will pass a 100-hr soak test in a 0.5% Rinso solution at 165 F. It will withstand immersion in 3% sodium hydroxide for 9 days at room temperature and still retain excellent gloss. Even after a drastic over-bake of 40 hr at 325 F., the film retains good color and high gloss.

Implement Enamel

	FORMULA No. 1	No. 2
	(Water Soluable – Alkyd)	
Linoleic Fatty Acid[1]	357	–
Linseed Fatty Acids	–	357
Trimethylol Propane	335	335
Amoco IPA-85	299	299
Amoco TMA	99	99

Procedure:

Charge fatty acid, trimethylol propane, and isophthalic acid to reaction kettle. Heat rapidly with agitation to 350 F (177 C) where esterfication begins. Heat over 3 hr period to 460 F 340–350 F (171–177 C) for 50–65 acid number and 20 s cure. Cool to 310 F (154 C). If hard resin modification is desired, blend in 5% resin for about 10 min. Thin to 75% solids in cosolvent.

	No. 1	No. 2
Resin Properties:		
Acid Number (solids)	50.4	62.9
Hydroxyl Number (solids)	230–240	
Visc. (Gardner-Holdt)	Z_6	Z_4
NVM %	75	75
Volatile	1/1 Butanol/Butyl **Cellosolve**[2]	
Cure at 200°C[3]	20 s	19 s
Color (Gardner)	6	7

---

[1] **Pamolyn 200**

[2] Ethylene Glycol Mono Butyl Ether

[3] Cure is the time required for the polymer to change from liquid to gelatinous state.

	No. 3	No. 4
	lb per 100 gal	
	(Chelate Cured – Alkyd)	*(Hard Resin Modified – Alkyd)*
WS-549 Alkyd (75% NVM)	326.0	–
WS-549 Hard Resin Modified (75% NVM)	–	328.0
Ammonium Hydroxide (28%)	15.0	20.0
Chelate[1]	15.0	3.0
Drier Accelerator[2]	1.0	0.5
Anti-Skin Agent[3]	1.0	1.0
Water	438.7	448.0
Titanium Dioxide[4]	219.8	221.0
Flow Agent[5]	0.5	0.5
Solution Properties:		
pH	8.0–8.5	8.0–8.5
PVC (%)	15	15
Spray Visc. No. 4 Ford Cup	40–50 s	40–50 s
Wt. % Solids	45.6	45.7

Drying Characteristics – $70 \pm 2°F(21 \pm °C)$, 50% RH

Set-to-Touch	15 min	3 min
Tack Free	45–60 min	15 min
Dry Hard	1.5–2.0 hr	45 min

Procedure:

Solubilize the resin with ammonium hydroxide. Dilute the mixture with ¾ of the water. Adjust pH to 8.0–8.5 if necessary. Add the drier accelerator, antiskin agent and flow agent. Mix thoroughly. Slowly stir in the TiO_2 while adding the remaining water. Grind the resulting paste in a

---

[1] Tyzor TE Organic Titanate
[2] Active 8
[3] Exkin No. 2
[4] TiPure R-900
[5] FC-430

pebble mill overnight to a 7 Hegman grind. When specific grind is reached, *stir* in the chelate. Put on the roller for 5 min for complete dispersion. Adjust with as much water as needed to spraying viscosity (40–50 s #4 Ford Cup).

Screw-Cap Coating

(Alkyd)

	lb	gal
Roller mill grind:		
Titanium Dioxide	284.0	8.74
Amberlac 80X (70% solids)	170.5	19.39
Hi-Flash Naphtha	113.7	15.64
Mix with:		
Amberlac 80X (70% solids)	227.0	25.81
Hi-Flash Naphtha	199.0	27.42
Pine Oil	14.2	1.84
Butyl Carbitol	8.5	1.07
Cobalt Naphthenate (6%)	0.7	0.09
Wt./gal		10.2 lb
Total Solids		57.3%
Pigment		50.5%
Binder		49.5%
Drier		0.015%

Bake for 10 min at 275°–300°F.

White Coating for Can Bodies

(Alkyd)

	lb	gal
Roller mill grind:		
Titanium Dioxide	271	8.4
Amberlac 292G (50% solids)	146	18.2
Mix with:		
Amberlac 292G (50% solids)	397	49.5

	lb	gal
Solvesso 150	175	23.9
Wt./gal		9.9 lb
Total Solids		55.0%
Pigment		50%
Vehicle		50%

Bake for 10 min at 300°F.

Metal Protective Coating

(Silicone)

GE SR-701 Resin (10% Solids)	28.0
Propellent 12/11 (50/50)	72.0

Procedure:

The **SR-701** is supplied ready for use in the aerosol container. Simply change the resin solution to an aerosol can and load the propellent.

Valve and Actuator Selection

The following valve has proven satisfactory for this application:

> From: Newman Green
> 151 Interstate Road
> Addison, Illinois
> Valve R-10-118 (.013 VT)
> Sprayhead 196-20-12

Application:

Clean the metal surface to remove any dirt, tar, and rust. Hold the can 9–10 in. from the substrate and spray in a slow sweeping motion.

Zinc Rich Primer

FORMULA NO. 1

(Epoxy)

Resin:

Epoxy Resin	42.40	⎫
Xylene	14.14	⎬ Precut
Cellosolve	14.14	⎭

Reduce with:

Butyl Alcohol	13.79	
Xylene	6.89	Visc.: 7 s No. 4 Zahn
MIBK	8.64	Wt. per gal: 8.13 lb

Hardener:

Uni-Rez 2115	19.10 ⎱ Same as **Uni-Rez 2415** which may
Xylene	9.17 ⎰ be used.

Reduce with:

Cellosolve	1.27
Xylene	6.37
Butyl Alcohol	6.21
MIBK	3.88

Visc.: 9 s No. 4 Zahn
Wt. per gal: 7.50 lb

Application:

This coating offers excellent long term protection to steel when top-coated, but may also be used as a semipermanent weld-through primer to protect steel during field erection and in weather exposed storage. Protection is by galvanic action and will prevent underfilm corrosion when the topcoat has been damaged.

Best results will be obtained when applied over sandblasted steel. Lesser adhesion and protection can be expected over wirebrushed or rusted surfaces.

Recommended dry film thickness is 1/2 to 3 mils.

Recommended topcoats are: epoxy-polyamide, epoxy ester, and solvent acrylic coatings.

No. 2

Resin:

Epoxy Resin	42.40	⎱
Xylene	14.14	⎬ Precut
Cellosolve	14.14	⎰

Reduce with:

Butyl Alcohol	13.79

| Xylene | 6.89 | Visc. 7 s No. 4 Zahn |
| MIBK | 8.64 | Wt. per gal: 8.13 lb |

Hardener:

Uni-Rez 2100	42.40	
Xylene	25.45	Same as **Uni-Rez 2401** which may
Cellosolve	2.83	be used.

Reduce with:

Butyl Alcohol	13.79
Xylene	6.89
MIBK	8.64

Visc.: 13 s No. 4 Zahn
Wt. per gal. 7.47 lb

Zinc Chromate Primer

(Epoxy – Polyamide)

		lb/100 gal
A	45-228 Zinc Yellow	308.0
	Nytal 300	53.0
	Bentone 27 } Prewet	4.5
	Ethyl Alcohol }	2.5
	Epotuf 38-501	257.0
	Xylol	88.0
	MIBK	72.0
B	Polyamide Hardener 37-618	152.0
	Xylol	81.0
	Cellosolve	20.0

Uses:

Industrial air-drying metal primer where maximum corrosion resistance is required.

Iron Oxide – Zinc Chromate Primer

(Epoxy)

		lb	gal
A	Pure Red Iron Oxide	100.0	2.33
	Celite 281	100.0	5.21
	Asbestine 3X	100.0	4.21
	Zinc Chromate	50.0	1.71
	Barytes #1	50.0	1.35
	Zinc Oxide	25.0	0.53
	Aluminum Stearate	3.0	0.15
	Epi-Rez 285	270.0	31.00
	Pine Oil	15.7	2.00
	Toluol	61.6	8.50
	Ethyl Cellosolve	179.0	23.01
	Visc. 68 KU; Wt./gal. 11.92	954.3	80.00

Grind in steel ball mill to 6–8 on Paint Club Scale

		lb	gal
B	Epi-Cure X-70-8515	77.2	9.90
	Toluol	24.2	3.34
	Xylol	48.6	6.76

Visc. 13 s (#4 Ford)
Wt./gal. 750

Composite Blend (allow 1 hr induction prior to application)

	lb	gal
A	954.3	80.00
B	150.0	20.00

Visc. (KU)	62
Wt./gal	11.04
Application	Spray or Brush
Schedule	Room Temperature Air Dry
PVC (%)	40.0
Total Solids (%)	58.3
Vehicle Solids (%)	32.0

Epoxy-Ester Zinc Chromate Spraying Primer

	lb/100 gal
45-228 Zinc Chromate	281.0
Nytal 300	50.0

	lb/100 gal
Epotuf 38-403	151.0
Xylol	145.0

Pebble mill to 6 N.S. fineness

Epotuf 38-403 (6403-50)	385.0
Lead Naphthenate (24%)	4.5
Cobalt Naphthenate (6%)	1.7
Exkin No. 2	1.0

Uses:
  Corrosion-inhibiting industrial primer.

Lead Silico Chromate Primer

(Epoxy)

Lead Silico Chromate		22.79	
Asbestine X Talc		25.77	
Bentone 27		.57	
Methanol		.57	
Epoxy Resin	18.80		
Xylene	6.27	Precut	31.34
Cellosolve	6.27		
Xylene		8.14	

Ball Mill 16 hr, wash mill with the following and use in batch:

Butyl Alcohol	2.24
Cellosolve	6.43 Visc.: 11 s No. 4 Zahn
Methyl Isobutyl Ketone	2.15 Wt. per gal: 12.56 lb

Hardener:

Uni-Rez 2115	8.45 ⎱ Same as **Uni-Rez 2415**
Xylol	3.62 ⎰ which may be used
Cellosolve	2.41 Visc.: 15 s No. 4 Zahn
Butyl Alcohol	1.21 Wt. per gal: 7.72 lb

Procedure:
  Blend resin and hardener together, let stand for 30 min induction period, filter before using. May be reduced as necessary for spraying.

Spray Reducer:

Xylene	60
MIBK	10
Cellosolve	30

Application:

For brush or spray application to steel and aluminum surfaces. Best results will be obtained when this primer is applied over sandblasted steel. Somewhat lesser results can be expected when applied over wirebrushed or rusted surfaces. Aluminum should be cleaned of all contamination by chemical or mechanical cleaning. Topcoats recommended are: epoxy ester and solvent acrylic coatings.

Baking Primers

FORMULA	No. 1	No. 2	No. 3
	(Water-Soluble — Epoxy)		
TMA Resin 408 (65% NVM)	330	330	330
Triethylamine	21	21	21
Water	390	390	390
Cobalt Naphthenate (6%)	3	3	3
Manganese Naphthenate (6%)	2	2	2
Anti-Skinning Agent	1	1	1
Red Iron Oxide[1]	100	100	100
Calcium Carbonate[2]	175	160	110
Strontium Chromate[3]	—	25	—
Basic Lead Silico Chromate[4]	—	—	100
Visc. #4 Ford Cup (s)	30–35	30–35	30–35
pH	8.0–8.5	8.0–8.5	8.0–8.5
NVM (Wt. %)	48	49	50
PVC (%)	30	30	30

[1] R-4098
[2] Duramite
[3] 1367
[4] M-50

Red Lead Metal Primer

FORMULA NO. 1

(Epoxy – Polyamide)

		lb/100 gal
A	Red Lead (97%)	900.0
	Nytal 300	80.0
	Superjet Lampblack	2.0
	Aluminum Stearate V	4.0
	Polyamide Hardener 37-618	140.0
	Solvesso 150	102.0
	Cellosolve	18.0
	Steel ball mill to 6 N.S. fineness	
B	**Epotuf 38-505 (6505-75)**	240.0
	Solvesso 150	67.0
	Cellosolve	111.0

High quality maintenance primer for structural steel and other metal surfaces.

NO. 2

(Epoxy-Ester-Brushing)

	lb/100 gal
Red Lead (97%)	920.0
1767 Talc	80.0
Soya Lecithin	2.0
Aluminum Stearate	4.0
Epotuf 38-406 (6406-60)	150.0
Mineral Spirits	150.0
Steel ball mill to 5+ N.S. fineness	
Epotuf 38-406 (6406-60)	232.0
Mineral Spirits	67.0
Cobalt Naphthenate (6%)	2.3
Lead Naphthenate (24%)	5.6
Calcium Naphthenate (4%)	2.3
Exkin No. 2	1.0

Uses:

Maintenance primer for steel structures.

No. 3

(Epoxy)

		lb
A	**Araldite 571 CX-80**	210.0
	Red Lead (97% fumed)	480.0
	Titanium Dioxide	30.0
	Red Iron Oxide	15.0
	Aluminum Stearate	4.5
	Calcium Silicate	235.0
	Xylene	198.0
	Diacetone Alcohol	99.0
	Flow Control Agent	10.5
B	**Araldite Hardener 820**	105.0
	Xylene	24.0
	n-Butanol	12.0

Red Lead/Red Iron Oxide General Purpose Flash Dry Primer

(Epoxy)

	lb	gal
Ball mill charge:		
Red Lead (97%)	114.1	1.51
Red Iron Oxide (**Mapico 516**)	114.1	2.66
Asbestine 3X	97.8	4.14
M-P-A (Xylene)	75.0	10.33
Eponol 53-L-32	257.2	29.89
Arochlor 5460	27.4	1.98
CKM-5254	27.4	2.79
Methyl Oxitol	52.5	6.52
Cyclo-Sol 53	26.2	3.59
Xylene	26.2	3.61
Mill rinse:		
Methyl Oxitol	126.0	15.67

	lb	gal
Cyclo-Sol 53	63.0	8.63
Xylene	63.0	8.68

Wt./gal (lb)	10.70
Solids Concentration, Calculated:	45.0% Wt.
Visc. (KU)	
Stormer Viscometer:	74
PVC (%)	
Calculated:	45
Dry to Handle	13 min

Procedure:

Charge the ball mill components into a steel ball mill and grind for 16–24 hr. Empty the mill. Add the mill rinse mixture and run the mill 30–60 min. Empty the mill and add this to the grind.

Application Instructions:

Spray Application: Thin 1.33 vol. of Primer base with 1.0 vol. of Methyl Oxitol, Cyclo-Sol 53, and Xylene (ratio 2/1/1 parts by weight). The presence of MPA-60 (Xylene) retards settling of pigment in the thinned formula on prolonged standing and improves the spray characteristics. The primer may be recoated within 1 to 2 hr and topcoated with either an amine cured Epon Resin type coating or an Epon Ester type coating after 16 hr.

Brush Application: Thin the Primer base to the desired brushing viscosity with the same solvent blend described under spray application. Flow the primer onto the surface with a minimum of rebrushing and do not "work" the coating excessively. The primer should not be recoated within 4 hr or topcoated within 24 hr.

Iron Oxide Primer

(Epoxy)

Epon 828	35.06	340.08
Epoxide 7	4.97	37.77
Mapico 387	13.69	590.72
Novacite 325	3.00	65.97
Asbestine 325	3.99	94.72
M-5 Cab-O-Sil	0.53	9.64

Glycerol	0.19	1.88
Beetle 216-8	2.23	18.96
Ethylene Glycol		
Monoethyl Ether	3.01	23.33
Uni-Rez 2341	18.43	151.14
Toluene	2.39	17.30
Ethylene Glycol		
Monoethyl Ether	9.17	71.07
n-Butanol	3.34	22.61

A high solids chemical and abrasion resistant epoxy primer using **Uni-Rez 2341** as the epoxy curing agent.

Application:
This coating may be applied to metal and concrete surfaces by brush roller, or spray up to .020 in thick.

Red Primer

(Epoxy)

A	**Araldite 471 X-75**	411.5
	Red Iron Oxide	93.0
	Barium Sulfate	138.8
	Calcium Silicate	45.9
	Cellosolve	59.9
	Solvesso 150	14.0
	MEK	13.0
	Flow Control Agent	2.0
	Antisag Agent	7.0
B	**CIBA Polyamide 815 X-70**	154.8
	MEK	55.9
	Solvesso 150	45.9
	Nonvolatile Content (%)	63.3
	Epoxy/Hardener Ratio	100/50
	Pigment/Binder Ratio	42/58
	Density, lb/gal, 25 C (77°F)	10.4

Automotive Flash Primer

(Epoxy)

	lb/100 gal
Iron Oxide (97% Fe_2O_3)	107.0
Lithopone (Medium Oil Absorption)	53.0
Lampblack	4.0
Calcium Silicate	107.0
Barium Sulfate	89.0
Xylene	65.0
Calcium Naphthenate (4%)	7.0
Triethylamine	2.0
Soya **Araldite 7098** ester (50% N.V.)	389.0

Ball mill grind to fineness of 6 N.S. Hegman
Gauge, then add:

Urea-Formaldehyde Resin	41.0
O-Amylphenol	2.0
VM&P Naphtha	64.0
High Flash Naphtha	64.0
Isopropyl Alcohol	9.5
Lead Naphthenate (24%)	7.0
Cobalt Naphthenate (6%)	2.0

Strontium Chromate Primer

(Epoxy)

	lb
Strontium Chromate	22.0
Rutile Titanium Dioxide	45.0
Magnesium Silicate	47.0
Diatomaceous Silica	28.0
Araldite 488 N-40	445.2
Methyl Ethyl Ketone	142.4
Methyl Isobutyl Ketone	42.1
Cellosolve	42.1
Xylene	42.1
Nonvolatile Content (%)	37.4

	lb
Pigment/Binder Ratio	45/55
PVC (%)	26
Visc. 25 C (77 F) Stormer Viscometer (KU)	60

Appliance and Metal Furniture Primer

(Epoxy)

	lb/100 gal
Titanium Dioxide	85.6
Soya **Araldite 7098** Ester (50% N.V.)	171.2
3 roll mill-fineness of 7-8 N.S., Hegman Gauge, then add:	
Soya **Araldite 7098** Ester (50% N.V.)	377.2
Melamine-formaldehyde resin	137.2
Xylene	85.2

Detergent-Resistant Appliance Primer

(Epoxy)

	lb
Rutile Titanium Dioxide	278.6
Lampblack	5.7
Araldite 9308 (40% N.V.)	192.5
Cellosolve Acetate	224.9
Xylene	224.9
Araldite DP-138 (50% N.V.)	82.9
Catalyst Solution*	7.1
Flow Control Agent	3.0

---

*25% sol'n. of salicylic acid in ethanol

Stainless Steel Industrial Maintenance Primer

(Epoxy)

		lb
A	Araldite 7071	160.0
	MIBK	53.4
	Cellosolve	53.4
	Xylene	53.4
B	CIBA Polyamide 800 CX-60	260.8
	Steel Pigment	219.2
	Aluminum Pigment Paste	14.9
	MIBK	71.9
	Xylene	75.6
	Cellosolve	17.0

Gray Baking Sanding Primer Surfacer

(Epoxy – Ester)

	lb/100 gal
Lithopone (medium oil)	58.0
Superjet Lampblack	1.5
ASP-100	100.0
No. 1 Barytes	100.0
Calcium Naphthenate (4%)	5.8
Titanox RA-50	100.0
Epotuf 38-403	121.5
Xylol	72.0
VM&P Naphtha	50.0

Steel ball mill to 6+ N.S. finess

Epotuf 38-403 (6403-50)	232.0
Xylol	33.0
Beckamine 21-510 (P-138-60)	51.0
Solvesso 100	21.0
VM&P Naphtha	72.0
Orthophen 85	1.7
Cobalt Naphthenate (6%)	1.7
Lead Naphthenate (24%)	0.85

Uses:
   Automotive primer surfacer and general industrial primer surfacer.

White Appliance Enamel

(Epoxy)

	lb
Rutile Titanium Dioxide	295.5
Araldite 7097 (100% N.V.)	236.4
Methyl Isobutyl Ketone	181.2
Toluene	155.8
Cellosolve Acetate	96.7
Melamine Resin (100% N.V.)	59.1
Flow Control Agent	3.0
Acid Catalyst	3.0

Note: Dissolve the melamine resin in an equal weight of toluene.

Epoxy White Gloss Enamel

FORMULA NO. 1

		lb
A	**Araldite 471 X-75**	393.7
	Titanium Dioxide	241.1
	MIBK	62.3
	Cellosolve	62.4
	Xylene	62.3
	Flow Control Agent	2.0
	Antisag Agent	3.0
B	**Araldite Hardener 835**	147.6
	Solvesso 150	29.5
Nonvolatile Content (%)		60.3
Epoxy/Hardener Ratio		100/50
Pigment/Binder Ratio		40/60
Density, lb/gal, 25°C (77°F)		10.0

No. 2

		lb
A	**Araldite 471 X-75**	276.5
	Rutile Titanium Dioxide	216.2
	Cellosolve	74.8
	˙Xylene	141.4
	Flow Control Agent	2.1
	Antisag Agent	4.2
B	**CIBA Polyamide 815 X-70**	218.3
	Cellosolve	35.3

Nonvolatile Content (%)	59.4
Epoxy/Hardener Ratio	100/72
Pigment/Binder Ratio	37/63
Density lb/gal, 25°C (77°F)	9.7
Visc. 25°C (77°) (KU)	64

Performance:

Substrate	Steel
Cure Schedule	7 days at 25°C. (77°F.) and 50% RH
Film Thickness, mils.	1.5
Adhesion	Excellent
Pencil Hardness	HB
Gloss, 60° geometry	100
Impact Resistance, in.-lb, Pass,	
direct	88
reverse	28

White Enamel

(Epoxy)

	lb
Rutile Titanium Dioxide	104.0
Araldite 488 E-32	473.1
Cellosolve	99.8
Methyl Isobutyl Ketone	99.8
Xylene	101.9

	lb
Nonvolatile Content (%)	29
Pigment Binder Ratio	40/60
PVC (%)	13.4
Visc., 25°C (77°F) Stormer Viscometer (KU)	94

Performance:

Cure Schedule, 25 C (77 F)	7 days
Substrate	Steel
Film Thickness	1.6 mils
Dry Time	
Set-to-touch	12 min
Cotton free	18 min
Tack free	35 min
Dry hard	3 hr
Specular Gloss (60°)	100+
Adhesion	Excellent
Impact Resistance, direct, in.-lb	64
Impact Resistance, reverse, in.-lb	20
Water Resistance, 25 C (77 F), 7 days	Loss of gloss
Xylene Resistance, 25 C (77 F), 2¼ hr	Softened to substrate

White Brushing Enamel

(Epoxy-Amine Adduct)

		lb/100 gal
A	**Titanox RA**	188.0
	Nytal 300	94.0
	Bentone 27	4.7
	Epotuf 38-505 (6505-75)	358.0
	Cellosolve	154.0
	Solvesso 150	18.0
	No. 840 Resin	2.8
	Pebble mill 7+ N.S. fineness	
B	**Epotuf 38-505 /6505-75)**	61.0
	Triethylene Tetramine	18.3
	Xylol	29.0

	lb/100 gal
Methyl Isobutyl Carbinol	45.0
Butyl Cellosolve	61.0

Procedure:

Mix **Epotuf** solution and solvents, then add amine slowly with good agitation. External cooling may be required. Allow to stand at least 24 hr before use.

Uses:

Maintenance enamel for chemical and petroleum plants, dairies, gas stations, etc.

White Metal Deco Enamel

(Epoxy)

	lb	gal
Rutile Titanium Dioxide	199.1	5.83
Araldite 497 C-55	434.5	48.27
Xylene	110.1	15.21
Diacetone Alcohol	110.1	14.08
o-Dichorobenzene	24.3	2.24
Melamine-Formaldehyde Resin (50% N.V.)	119.2	14.37

Machinery Enamel

(Epoxy)

	lb/100 gal
Titanium Dioxide	100.0
Mapico Yellow Orange	10.0
Superjet Lampblack	5.0
Epotuf 38-401	50.0
Hi-Flash Naphtha	50.0

Steel ball mill to 7 / N.S. finess

Epotuf 38-401 (6401-50)	604.0
Hi-Flash Naphtha	48.0
Cobalt Naphthenate (6%)	2.2
Lead Naphthenate (24%)	5.5
ASA	1.0

Gray Enamel

(Epoxy – Ester)

TMA Resin WS-265 (80% NVM)	398.0
Butoxyethanol	57.0
Cymel-301	56.0
Dimethylethanolamine	19.0
Water	800.0
Mogul A	2.0
TiPure R-900	73.0

Procedure:

Grind the materials listed in a pebble mill to a #7 Hegman grind. Reduce with additional water to spray viscosity of 35–40 s #4 Ford cup.

Initial pH	7.5–8.5
Pigment/Binder Ratio	0.2/1.0
Melamine/Alkyd Ratio	15/85

Epoxy Green Enamel

	lb	gal
Base:		
Vanoxy 201-X-75	310.4	34.10
Rayox R-88	123.4	3.59
Chromium Oxide Green X-1134 C.P.	139.8	3.30
Ramapo Green BGP-501-D	14.0	0.99
M-P-A Xylene	2.4	0.33
Beetle 216-8	12.5	1.47
Modaflow	1.9	0.23
Methyl Isobutyl Ketone	39.8	5.99
Curing Agent:		
Vanamid 315-X-70	184.6	23.67
Xylene	55.3	7.73
Vansolve EE	122.6	15.78
Vansolve EB	21.1	2.82

Total Nonvolatile (%)	63
Pigment/Binder, Wt. (%) Ratio	43.5/56.5

Wt. per gal, lb	10.3
Visc., Stormer (KU)	69

Procedure:

Base — Disperse the pigments and **M-P-A Xylene** in a suitable portion of the **Vanoxy 201-X-75** and available solvent using a 3 roll mill. Let down with the remaining **Vanoxy 201-X-75**, solvent **Beetle 216-8** and **Modaflow**.

Curing Agent — Charge the **Vanamid 315-X-70** to suitable container. Under constant agitation add the solvents and mix thoroughly.

Package the base component and the curing agent component separately. Mix just prior to use. Allow one hour induction time before applying.

Clear Epoxy Coating

(Amine Converted)

Araldite 571 KX-75	46.3
M-P-A 60 (Xylene)	10.4
MIBK	0.3
Cellosolve	7.8
Beetle 216-8 (60%)	1.8
Curing Agent:	
Araldite DP-123	25.0

Procedure:

Grind **Araldite 571-KX-75** and **M-P-A 60** (Xylene) on a 3 roll mill, a pebble mill or similar suitable equipment (at about 110°F). Let down with the **Beetle 216-8** and remaining solvents.

Application:

For industrial maintenance and metal protection, this room temperature curing epoxy provides outstanding chemical resistance.

Epoxy Clear Film

Formula	No. 1 (High Bake)	No. 2 (Low Bake)
TMA Resin 408 (80% NVM)	225	255
Triethylamine	18	18
Amino Crosslinker I[1]	–	45
Amino Crosslinker II[2]	36	–
Manganese Naphthanate (6%)	2	2
Cobalt Naphthanate (6%)	–	–
Water	169	160
Initial pH	7.5–8.0	7.5–8.0
Wt. % NVM	50	50
Melamine/Resin Ratio	15/85	15/85

Performance (Draw-down 1.5 mils wet)

Substrate	Cold Rolled Steel	
Cure Cycle, min/F	30/350	30/250
Film Thickness, mils	1.0	1.0
Sward Hardness	78	48
Pencil Hardness	H	F
Crosscut Adhesion (% Pass)	100	100
1/8" Conical Band (% Pass)	100	100
Impact Direct in lb	70	80
Impact Reverse, in lb	30	80

[1] MM-83
[2] Cymel 301

Black Aerosol Enamel

(Epoxy)

	lb	gal
Base:		
Superba Beads Special	20.0	1.37
Litharge	4.0	0.05
Soya Lecithin	8.0	1.00

	lb	gal
Post-4	8.7	1.00
Ten Cem Copper (6%)	1.1	0.15
Epi-Tex 183	200.0	25.00
Toluene	99.0	13.67

Steel ball mill 45–48 hr. Check grind using thin film draw-down, then add:

	lb	gal
Epi-Tex 183	200.0	25.00

Grind 1 hr and add:

	lb	gal
Epi-Tex 183	252.0	32.0
Lead Naphthenate (24%)	3.0	0.31
Cobalt Naphthenate (6%)	2.4	0.30
Exkin #2	1.2	0.15

Reduction:

	lb	gal
Base	356.0	44.5
Toluene	396.0	55.0
Cellosolve Acetate	4.0	0.5

Reduction Visc. — 11–13 s (Ford Cup #4).

Can Loading

	lb	gal
Reduced Paint	57.2	50.7
Propellant "P"	42.8	49.3

Visc. (KU)	88
Wt./gal	8.00
PVC (%)	3.58
Total Solids (%)	46.4
Vehicle Solids (%)	44.2

Gray Spraying Finish

(Epoxy — Amine)

	lb/100 gal
Titanox RA	125.0
Superjet Lampblack	3.5

	lb/100 gal
Imperial IAF Compound	2.0
Nytal 300	100.0
Blanc Fixe No. 333	150.0
Bentone 27	7.0
Epotuf 38-501 (6501-75)	225.0
Xylol	112.0
MIBK	60.0
Cellosolve	14.0

Steel ball mill 6÷ N.S. fineness

Epotuf 38-501	110.0
Beckamine 21-510	9.0
MEK	20.0

Catalyst:

Diethylene Triamine	16.0
Butyl Alcohol	28.0
Xylol	28.0

Reducer:

MIBK	35.0
Xylol	35.0

Uses:

Lining for tankcars and trucks carrying jet fuel, general industrial chemical-resistant coating.

Unpigmented Can Coating

FORMULA NO. 1

(Epoxy)

	lb
Araldite 488 N-40	528.9
Methyl Ethyl Ketone	25.7
Cellosolve Acetate	10.5
Toluene	8.6
Araldite DP-139 (50% N.V.)	180.2
Catalyst Solution*	45.3

Nonvolatile Content (%)	38.4
Epoxy/Phenolic Resin Ratio (solids)	70/30
% Catalyst (on total solids)	1.5
Recommended Cure Schedule	30 min at 177 C (350 F)

*10% solution of conc. H_3PO_4 in Cellosolve acetate

No. 2

(Epoxy)

	lb
Araldite 9307 (40% N.V.)	510.5
Cellosolve Acetate	127.6
Toluene	127.6
Urea-Formaldehyde Resin (60% N.V.)	85.1

Nonvolatile Content (%)	30
Visc., No. 4 Ford Cup, 25 C. (77 F.)	43 s

Performance:

Cure Schedule	30 min at 180°C. (356°F)
Substrate	Steel
Film Thickness, mil	0.2
Adhesion	Excellent
Pasteurization Test, 90 min at 82 C (180 F)	No effect
Boiling Water Test, 90 min at 100 C (212 F)	No effect
Sterilization Test, 90 min at 116 C (240 F), 10 psi	No effect
Acetone Resistance, 5 min	Softened

No. 3

(Epoxy)

	lb
Araldite 9307 (40% N.V.)	474.7
Cellosolve Acetate	112.4

	lb
Toluene	112.4
Araldite DP-139 (50% N.V.)	126.7
Catalyst Sol'n.*	18.9

Nonvolatile Content	30
Visc., No. 4 Ford Cup, 25 C. (77 F.)	30 s

Performance:

Cure Schedule	20 min at 200°C.(392°F)
Substrate	Steel
Film Thickness, mil	0.2
Adhesion	Excellent
Impact Resistance, direct (in lb)	160
Impact Resistance, reverse (in lb)	160
Flexibility, cylindrical mandrel (1/8 in)	Pass
Boiling Water Resistance (4 hr)	No effect
Boiling 20% NaOH Resistance (4 hr)	Yellow discoloration
Boiling 2% Acetic Acid Resistance, 1 hr	No effect
10% HCl Resistance, 25 C (77 F), 6 days	No. 6 blisters, medium

---

*10% sol'n. of conc. H_3PO_4 in **Cellosolve**

No. 4

(Epoxy)

	lb
Araldite 7097 100% NV	183.1
Cellosolve Acetate	245.3
Toluene	244.9
Araldite DP-139 50% NV	122.0
Catalyst Sol'n.*	18.3

---

*10% sol'n. of conc. H_3PO_4 in **Cellosolve**

Can Coating and Drum Lining

(Epoxy)

		lb	gal
A	**Resypox 1577**	222.2	23.26
	Neosol Solvent	83.3	12.34
	MEK	50.0	7.45
	Pine Oil	33.3	4.24
	Toluene	166.6	23.0
B	**Beckamine P-196**	158.7	18.50
	Neosol Solvent	19.8	2.94
	MEK	11.9	1.77
	Pine Oil	7.9	1.01
	Toluene	39.7	5.48

Reduce to #4 Ford Cup viscosity of 20 s of diacetone alcohol/xylene, 1:1 by weight.

Bake: 20 min @ 300°F.
                     or
               6 min @ 350°F.

For pigmentation, add 0.2% dye and bake 20 min @ 385°F for complete cure.

Unpigmented Drum Lining

(Epoxy)

Araldite 485 E-50	366.6
Toluene	246.4
Cellosolve Acetate	61.1
Araldite DP-139 (50% N.V.)	122.2
Catalyst Sol'n.*	18.3

---

*10% sol'n. of conc. H_3PO_4 in Cellosolve

Clear Interior Drum Lining

(Epoxy – Phenolic)

	lb/100 gal
Epotuf 38-503	338.0
Varcum 29-153	177.0
Xylol	124.0
Methyl Isobutyl Ketone	155.0
Cellosolve Acetate	16.0
Beckamine 21-510	5.6

Uses:

Lining for tanks, drums, cans, piping carrying corrosive chemicals and solvents.

Unpigmented Tank Lining

(Epoxy)

Araldite 597 KT-55	416.0
Toluene	92.4
Methyl Ethyl Ketone	46.5
Diacetone Alcohol	46.5
Araldite DP-139 (50% N.V.)	152.6
Flow Control Agent	3.1
Catalyst Sol'n.*	45.8

---

*10% sol'n. of conc. H_3PO_4 in Cellosolve

High Solids Tank Liner

(Epoxy)

		lb	gal
A	Basic Lead Silico Chromate (M-50)	100.0	2.94
	S.F. Magnesium Silicate	75.0	3.18
	Celite 281	25.0	1.30
	Thixcin R	20.0	2.43
	Epi-Rez 509	465.0	48.00

Procedure:

Premix above on Cowles Dissolver to 100 F; then grind 2 passes on a 3 roll mill. Add following in let down:

Mobilsol 44	77.2	8.82

Visc. off scale Wt./gal 11.45

B	**Epi-Cure 890**	207	29.00
	Xylol	31.1	4.33

Visc. 12 s #4 Ford Cup — Wt./gal 7.19

Composite
  Blend:

A	762.2	66.67
B	238.1	33.33

Visc., A & B (KU)	85
Wt./gal A & B	10.00
PVC (%)	9
VNV (%)	96
TNV (%)	97
Pot Life in pint sample	2.5 hr @ 75°F
Volume Solids (%)	95.6

Note: It is necessary to follow the manufacturing procedure listed above since a temperature of 100 F must be reached in order to obtain maximum puff from the thixotrope (**Thixcin R**). It should also be noted that too high a temperature (120 F+) will cause this thixotrope to seed out.

Chemical Resistant Metal Coating

Resypox 1577 (CAX)	374.1	43.5
Methylon 75108	69.3	6.93
Silicone Resin SR-82	4.7	0.53
Phosphoric Acid (85%)	5.0	0.35
n-Butanol	37.6	5.56
Cellosolve Acetate	157.6	19.43
Xylene	157.6	21.13
Yellow Oxide	97.43	

System Solids = 60.4%
Reduce to 30–32% N.V. with **Cellosolve Acetate**/Xylene mix at 1 : 1
Bake 10 min @ 400 F

White Roller Coat Tube Enamel

FORMULA NO. 1

(Epoxy)

Designed for coating collapsible tubes or other metallic objects requiring a hard, flexible coating with good adhesion. This coating is based on a single package system consisting of a DCFA epoxy ester and a melamine formaldehyde resin.

	lb
Rutile Titanium Dioxide	311.0
DCFA Epoxy Ester, (50% N.V.)	180.0
High Boiling Aromatic Solvent	26.0

Grind on a roller mill and add:

DCFA Epoxy Ester	380.0
Melamine Formaldehyde Resin (50%)	62.0
High Boiling Aromatic Solvent	48.4
Butanol	26.8
Rare Earth Napthenate (4%)	2.1

NO. 2

(Epoxy)

Rutile Titanium Dioxide	212.1
Pigment Suspension Agent	1.5
Araldite 488 N-40	662.8
Urea-Formaldehyde Resin (60% N.V.)	88.5
Flow Control Agent	3.1

Note: An acid catalyst, at about 0.5% based on vehicle solids, can be used to accelerate the cure.

Nonvolatile Content (%)	55
Pigment/Binder Ratio	40/60
Epoxy/UF Resin Ratio (solids)	83/17
Recommended Cure Schedule	20 min at 191 C (375 F)

Red Lead-Iron Oxide Mastic Coating

(Epoxy – Amine)

		lb/100 gal
A	No. 1 Barytes	180.0
	No. 1767 Talc	180.0
	Mapico No. 516 Red Iron Oxide	49.0
	Red Lead (97%)	164.0
	Bentone 27	5.6
	Soya Lecithin	3.0
	Epotuf 38-507 (6507-75)	149.0
	Epotuf 38-504 (6504-55)	205.0
	Toluol	123.0
	Methyl Isobutyl Ketone	161.0
	Steel ball mill 6 ÷ N.S. fineness	
B	Triethylene Tetramine	8.5
	Butyl Alcohol	8.5

Uses:

One-coat lining for interior steel structures, such as ship holds, railroad tankcars, etc.

Resin Coating

(Epoxy – Ester)

Tall Oil Fatty Acid [1]	424
Epoxy Resin [2]	520
Amoco TMA	68
Water of reaction	12

Resin Properties:

Acid Number	50–55
Visc., Gardner-Holdt	Z_8–Z_{10}

NVM (%)	80±1
Volatile	Butoxyethanol
Cure, s at 200 C	35–40
Color, Gardner	6–7

Procedure:

Charge the fatty acid and epoxy to a kettle equipped with inert gas sparge, agitator, thermometer, and short air-cooled condenser. Heat to melt the charge and begin agitation as soon as possible. React the epoxy and fatty acid to a top temperature of about 230 C (450 F) and hold for an acid number of 10–12.

Cool the reaction mixture to 170–180 C (340–360 F), charge the TMA and hold at that temperature for final properties. Cool the resin, reduce to 80% solids in butoxyethanol and filter.

---

[1] Acintol FA-1
[2] Epon 1004

Aluminum Paint

(Tri-Ester)

	lb	gal
MD 515 Aluminum Paste	180.0	14.76
Mineral Spirits	241.0	37.00
Chemacoil TA-303 Resin (80%)	396.0	52.00
Cobalt Drier (6%)	2.5	0.30
Zirconium Drier (6%)	5.5	0.76
Exkin #2 Antiskinning Agent	0.8	0.10

A ready-mixed aluminum paint possessing better package stability than an identical formulation using a 33-gallon oil length pure tung oil-phenolic spar varnish as the vehicle. After 8 months in a can 1/6 full, formulation XC-3200A was in good physical condition and still showed good leafing, whereas the varnish-based formulation had gelled. The two paints also were applied to bare steel panels, air dried, scribed, and then exposed vertically facing east in an industrial area. After 7 months the paint based on **Chemacoil TA-303** had better gloss, a better metallic sheen, and better color with less dirt collection. Both paints showed good adhesion and excellent rust resistance.

Leafing aids may be incorporated as desired in this formulation. If further protection against gas formation in sealed containers is required, silica gel may be included in the formulation.

Interior Mill White Gloss Enamel

(Triamino – Ester)

	lb	gal
Titanox RA-NC Titanium Dioxide	250.0	7.50
XX601 Zinc Oxide	100.0	2.13
Aluminum Stearate	1.0	0.12
Bentone 38 Gel (10% in odorless		
mineral spirits)	20.0	2.85
Chemacoil TA-303 Resin (90%)	400.0	52.75
Odorless Mineral Spirits	201.6	32.25
Cobalt Drier (6%)	2.0	0.25
Zirconium Drier (6%)	5.0	0.63
Antiskinning Agent	2.0	0.28
Maxvar 2503 Additive (55%)	21.6	3.00

Visc., – Fresh (KU)	90
Overnight (KU)	99
90 days (KU)	101
PVC (%)	19.2
Vehicle Nonvolatile (%)	49.4
Wt./gal (lb)	9.85

This formulation is designed for use on interior surfaces where mildew control is required, particularly when mercury compounds cannot be used or do not provide sufficiently lasting protection.

Coil Coating – Resistant Siding Type

(Polyurethane)

	Wt.	Solids	Equivalent	Group
Multron R-12A	209	209.0	0.62	-OH
TiO_2, R-KB-2	242	242.0		
EAB 381-2 (10% in				
ethylglycol acetate)	35	3.5		

	Wt.	Solids	Equivalent	Group
Modaflow (10% in ethylglycol acetate)	7	0.7		
Ethyl Glycol Acetate	140			
Mondur HCB	367	146.8	0.65	-NCO
Solids (%)	60			
NCO/OH	1.05			
PVC (%)	15			

Curing Schedules:

60–90 s @ 500 F
   5 min @ 400 F
  10 min @ 350 F
  30 min @ 300 F

This coating system is characterized by a combination of good weather and chemical resistance. It is recommended for evaluation as an industrial siding enamel.

Polyester White Coil Coating Enamel

	lb/100 gal		
FORMULA	No. 1	No. 2	No. 3
	(Flat)	*(Semi)*	*(High)*
	611	610	609
	558	557	553
Unitane OR 640-A Titanium Dioxide	242.1	297.0	336.6
Asbestine 3X Mag. Sil. Fibrous	48.4	25.6	—
Fibrene C400 Talc. Fibrous Acicular	193.7	102.5	—
Cyplex Resin 1476-8 (65%)	111.5	110.2	172.4
Solvesso 150 Solvent	148.1	92.6	13.7
Isophorone	37.5	23.3	—
Disperse [1] and add:			
Cyplex Resin 1476-8 (65%)	203.8	299.0	316.3
Cymel Hexamethoxymethylmelamine	36.0	46.9	56.1

	lb/100 gal		
	No. 1	No. 2	No. 3
	(Flat)	*(Semi)*	*(High)*
Catalyst 1010	2.3	3.1	3.8
Isophorone	5.2	16.5	38.3
Butanol	40.3	36.9	33.9
Solvesso 150 Solvent	83.9	82.3	123.4
Total Solids (% by Wt.)	63.0	65.0	65.0
by vol.	42.3	46.3	48.2
PVC (%)	41.3	30.6	20.4
Binder (% by Wt.) solids basis,			
Cyplex Resin 1476-8	85.0	85.0	85.0
Cymel Hexamethoxymethylmelamine	15.0	15.0	15.0
Catalyst 1010 (% on total resin solids)	1.0	1.0	1.0

Anticratering agents such as 3% 1/2 s cellulose acetate-butyrate may be helpful for flow control in high gloss formulations based on **Cyplex Resin 1476-8**. Usage is based on total resin solids.

---

[1] Disperse flat and semigloss formulations with pebble mill; high gloss with roller mill.

No. 4

(High Gloss)

	lb/100 gal
Unitane Or-650 Nonchalking Titanium Dioxide	334.2
Cyplex 1544 Resin (65%)	171.4
Disperse on 3-roll mill and add:	
Cyplex 1544 Resin	392.7
Cymel 303 Hexamethoxymethylmelamine (100%)	41.3
Cymel 247-10 Resin (60%)	21.1
Catalyst 1010	4.2
Silicone L5310 Resin	0.4

	lb/100 gal
Ethylene Glycol Monoether Acetate	52.8
Diacetone Alcohol	3.4
Isophorone	6.9
Ethylene Glycol Monobutyl Ether	8.4
Solvesso 150 Solvent	45.2
n-Butanol	38.1

White High Gloss Electrocoating

(Polyester)

	lb
Cyplex 1600 Resin	306.1
Cymel 1116 Cross-Linking Agent	76.4
Triethylamine	18.3
Unitane OR-600 Titanium Dioxide	143
Deionized Water	354

Primer for Galvanized

(Polyvinyl Chloride)

	lb	gal
Santicizer 8	30.0	3.03
Solvesso 100	133.0	18.30
Butyl Acetate	133.0	18.30
Cellosolve Acetate	133.0	16.44
Vinoflex MP 400 (medium)	250.0	24.14
MPA 60 (xylene)	15.0	2.06
Titanox CL	75.0	2.19
Nytal 300	100.0	4.21
Zinc Oxide XX601	20.0	0.43
Mineral Spirits	235.0	36.15

PVC (%)	18
Nonvolatile (%)	43
Nonvolatile by vol. (%)	28
Wt./gal (lb)	8.97
Vis. (KU)	69
Visc., Sec. DIN #4 Cup	38
(20% Reduction by vol. with Toluene)	

Red Iron Oxide Primer

(Latex)

		lb	gal
1.	**XC-4011** Resin	182.00	21.41
2.	**Cymel 1123** Cross-Linking Agent	53.8	5.55
3.	Diisopropanolamine	20.5	2.41
4.	Kroma Iron Oxide Red RO-3097	75.3	1.85
5.	Oncor F-31	32.3	0.75
6.	Deionized Water	332.3	39.93
7.	**XC-4011** Resin	227.0	26.71
8.	Diisopropanolamine	11.8	1.39

Procedure:

XC-4011 resin (1) is blended with **Cymel 1123** cross-linking agent (2), diisopropanolamine (3) and the pigments (4 & 5). To this blend the deionized water (6) is added under high speed agitation. The resulting paste is ground on a ball mill for 24 hr. The ground mixture is let down with the blend of **XC-4011** (7) and diisopropanolamine (8). The 50% concentrate is diluted to the final bath concentration of 10%.

Dark Grey Single Coat or Primer

(Latex)

	lb	gal
XC-4011 Resin	402.8	47.39
Cymel 1116 Cross-Linking Agent	75.1	7.95
Diisopropanolamine	31.9	3.75
Unitane OR-600 Titanium Dioxide	63.7	1.86
Chrome Yellow 40-4500	3.4	0.07
Raven 50 Carbon Black	8.0	0.55
Deionized Water	319.8	38.43

Beige Single Coat

(Latex Salt Spray Resistance)

	lb	gal
XC-4011 Resin	379.0	44.59
Cymel 1116 Cross-Linking Agent	71.2	7.53
Diisopropanolamine	30.0	3.53
Unitane OR-600 Titanium Dioxide	67.9	2.00
Chrome Yellow 40-4500	5.5	0.11
Raven 50 Carbon Black	0.46	0.03
Aluminum Silicate		
ASP 200	32.4	1.51
Deionized Water	338.6	40.70

Procedure:

Half the amount of the **XC-4011** resin is blended with **Cymel 1116** cross-linking agent, the diisopropanolamine and the pigments. The resulting paste is then ground on a three roll mill and let down with the remainder of the **XC-4011** resin. Deionized water is added slowly under high speed agitation to the resulting paste. This concentrate, 50% solids, is further diluted to the final bath solids of 10%.

Latex Maintenance Enamels

Formula	No. 1	No. 2	No. 3
	(Light Grey)	*(White)*	*(Grey)*
Aromatic High Flash Naphtha	210.0	220.0	210.0
Marbon 9200 MV	150.0	150.0	150.0
	Mix to clear solution		
Chlorinated Paraffin (50%)	75.0	75.0	75.0
Raw Linseed Oil	15.0	15.0	15.0
Zinc Oxide	20.0	20.0	20.0
Aluminum Silicate	40.0	–	65.0
Titanium Dioxide-Rutile	150.0	250.0	100.0
Suspension Additive*	1.0	1.0	1.0
Denatured Alcohol*	1.0	1.0	1.0
Anti-Floating Additive	1.0	–	1.0
Lecithin	2.0	2.0	2.0

	No. 1 (Light Grey)	No. 2 (White)	No. 3 (Grey)
(*Prewet)	Mix to smooth paste and grind on 3-roll mill to fineness of 7 N.S.		
Mineral Spirits	210.0	220.0	210.0
Cobalt Tallate (6%)	4.0	4.0	4.0
Lampblack in Oil	6.0	–	7.0
Yellow Iron Oxide in Oil	2.0	–	–

Procedure:
  Add ingredients in the order shown.

	No. 1	No. 2	No. 3
Total Solids (%)	52.6	54.0	51.1
PVC (%)	21.5	22.8	19.8
Wt. per gal (lb)	9.3	9.7	9.0
Visc. (KU)	77	78	76
Gloss 60°	90	92	92

Drying time:			
Set-to-Touch	40 min	40 min	40 min
Tack free	2.5 hr	2.5 hr	2.5 hr
Dry hard	4.0 hr	4.0 hr	4.0 hr
Dry through	4.0 hr	4.0 hr	4.0 hr

	No. 4 (Red)	No. 5 (Green)
Aromatic High Flash Naphtha	220.0	220.0
Marbon 9200 MV	150.0	150.0
	Mix to clear solution	
Chlorinated Paraffin (50%)	75.0	75.0
Raw Linseed Oil	15.0	15.0
Zinc Oxide	20.0	20.0
Aluminum Silicate	65.0	65.0
Red Iron Oxide	125.0	–
Chrome Green (medium)	–	100.0
Suspension Additive*	1.0	1.0

	No. 4 *(Red)*	No. 5 *(Green)*
Denatured Alcohol*	1.0	1.0
Lecithin	2.0	2.0
(*Prewet)	Mix to smooth paste and grind on 3-roll mill to fineness of 7 N.S.	
Mineral Spirits (KB40+)	220.0	220.0
Cobalt Tallate (6%)	4.0	4.0

Procedure:
  Add in the order as shown

	No. 4	No. 5
Total solids (%)	50.9	49.4
PVC (%)	20.5	20.0
Wt. per gal (lb)	9.2	9.0
Visc. (KU)	79	79
Gloss 60°	89	83

Drying time:

	No. 4	No. 5
Set-to-Touch	40 min	40 min
Tack free	2.5 hr	2.5 hr
Dry hard	4.0 hr	4.0 hr
Dry through	4.0 hr	4.0 hr

Exterior Latex Metal Paint

	lb	gal
Premix:		
Water	166.6	20.00
Triton X-100 Surfactant	5.0	0.56
Victawet 35-B Surfactant	2.5	0.26
Potassium Tripolyphosphate	3.0	—
Increase speed and add slowly:		
Chemacoil TA-100	194.0	24.83

	lb	gal
Adjust speed to disperse pigments:		
XX601 Zinc Oxide	100.0	2.14
Titanox RA-50 Titanium Dioxide	200.0	5.72
Nytal 300 Talc	100.0	4.21
Ethylene Glycol	40.0	4.32
Guardsan Preservative	1.5	0.24
Colloid 581-B Defoamer	2.0	0.30
Slow to mixing speed:		
Cellosize QP-4400 (3%) Thickener	180.0	21.61
Dow 300 Butadiene-Styrene Latex	37.0	4.35
Super-Cobalt Drier	1.8	0.22
Water or Thickener Sol'n.	94.0	11.28

Visc. — Fresh (KU)	77
Overnight (KU)	78
10 months (KU)	82
PVC (%)	31%

Chemacoil TA-100 = 92% of binder solids

Interior Latex Flat Wall Paint for Machine and Tube Colorants

Water	394.0
Advacide 60	.15
KTPP	1.2
Tamol 731	7.0
Igepal CO 630	3.0
Titanox RA-46	200.0
Hydrite Flat	100.0
Drikalite	215.0
Defoamer	3.0
Parco 37-C-55	245.0
Ethylene Glycol	25.0
Natrosol 250 MR	5.0
Carbitol Acetate	9.0
Yield (gal)	100.1
PVC (%)	59
Solids (%)	53

Visc. (KU)		90–95
Wt./gal (lb)		12.1

Latex Aluminum Paint

FORMULA	No. 1 (Bronze Metallic)		No. 2 (Gold Metallic)		No. 3 (Copper Metallic)	
	lb	gal	lb	gal	lb	gal
Mineral Spirits	336.0	50.9	336.0	50.9	336.0	50.9
Marbon 1100T MV	150.0	17.6	150.0	17.6	150.0	17.6

Mix to clear solution

Chlorinated Paraffin (50%)	75.0	7.2	75.0	7.2	75.0	7.2
Raw Linseed Oil	37.5	4.8	37.5	4.8	37.5	4.8
Dipentine	20.0	2.8	20.0	2.8	20.0	2.8
Cobalt Tallate (6%)	2.5	0.3	2.5	0.3	2.5	0.3
Zirconium Octotate (6%)	1.0	0.1	1.0	0.1	1.0	0.1
Silicone Fluid (1% in M. S.)	0.2	0.03	0.2	0.03	0.2	0.03
Lecithin	4.0	0.5	4.0	0.5	4.0	0.5

Mix until dissolved. Add the solution to the aluminum paste slowly with constant slow stirring.

Aluminum Paste	125.0	10.2	125.0	10.2	125.0	10.2
Aluminum Stabilizer	6.7	0.6	6.7	0.6	6.7	0.6
Burnt Sienna in Oil	40.0	3.1	–	–	–	–
Yellow Iron Oxide in Oil	–	–	66.0	6.3	–	–
Red Iron Oxide in Oil	–	–	–	–	40.0	2.9

Procedure:
    Add ingredients in the order as shown.

Latex Aluminum Paint (cont.)

Formula	No. 1 (Bronze Metallic)		No. 2 (Gold Metallic)		No. 3 (Copper Metallic)	
	lb	gal	lb	gal	lb	gal
Total Solids (%)		49.5		51.0		49.5
Wt. per gal (lb)		8.1		8.1		8.1
Visc. (KU)		71		72		72
Gloss 60°		51		48		45
Drying time:						
Set-to-Touch		25 min		25 min		25 min
Tack Free		3.5 hr		3.5 hr		3.5 hr
Dry Hard		15 hr		15 hr		15 hr
Dry Through		24 hr		24 hr		24 hr

	No. 4 (Heat Resistant)	No. 5 (Green Metallic)	No. 6 (Blue Metallic)
	lb	lb	lb
Xylol	192.0	–	–
Marbon 9200 MV	130.0	–	–
Mineral Spirits	–	336.0	336.0
Marbon 1100T MV	–	150.0	150.0
Mix to clear solution			
Silicone Resin	24.1	–	–
VM&P Naphtha	200.0	–	–
Chlorinated Paraffin (50%)	–	75.0	75.0
Raw Linseed Oil	–	37.5	37.5
Dipentine	–	20.0	20.0
Cobalt Tallate (6%)	–	2.5	2.5
Zirconium Octoate (6%)	–	1.0	1.0
Silicone Fluid (1% in M. S.)	–	0.2	0.2
Lecithin	–	4.0	4.0

Mix until dissolved. Add the solution to the aluminum paste slowly, with constant slow stirring.

	No. 4 (Heat Resistant)	No. 5 (Green Metallic)	No. 6 (Blue Metallic)
	lb	lb	lb
Aluminum Paste	290.0	—	—
Aluminum Paste	—	125.0	125.0
Aluminum Stabilizer	—	6.7	6.7
Phthalocyanine Green in Oil	—	33.0	—
Phthalocyanine Blue in Oil	—	—	25.0

Procedure:
  Add ingredients in the order as shown.

	No.4	No.5	No.6
Total Solids (%)	40.9	48.5	47.9
Wt. per gal (lb)	8.4	8.0	8.0
Visc. (KU)	69	71	72
Gloss 60°	51	52	52

Drying time:

	No.4	No.5	No.6
Set-to-Touch	8 min	25 min	25 min
Tack Free	10 min	3.5 hr	3.5 hr
Dry Hard	75 min	15 hr	15 hr
Dry Through		24 hr	24 hr

Zinc Chromate, Corrosion Inhibiting Primer

(Acrylic)

	lb	gal
Water	217.0	26.05
TKPP	1.0	0.10
Ethylene Glycol	30.0	3.23
Mineral Spirits	15.0	2.35
Cosan PMA - 30	0.5	—
Cellozise QP-4400	2.0	0.20
Texanol	10.0	1.12
Titanox RAN C	65.0	1.90
Zinc Chromate 45-228	125.0	4.27

	lb	gal
Mica 325 Mesh WG	25.0	1.06
Barytes X-5-R	50.0	1.34
Cabolite P-4	90.0	3.71
Deefo 97-2	1.0	—
Acronal 290 D	185.0	21.51

Disperse on high speed disperser.

	lb	gal
25% Phtalopal SEB-M*	40.0	4.60
Deefo 97-2	1.0	—
Acronal 290 D	350.0	40.69
NH_4OH (28%)	2.0	0.10

*25% Phtalopal SEB-M

Phtalopal SEB	250.0
Morpholine	70.0
Water	680.0

PVC (%)	28.0
Nonvolatile (%)	52.5
Nonvolatile by vol. (%)	38.7
Wt./gal (lb)	10.77
Visc. (KU @ 25°C)	82±3

Maintenance Primer

Formula No. 1

(White — Acrylic)

	lb	gal
Water	168.0	20.17
Latekoll D (8%)	50.0	5.92
Pigment Disperser A (30%)	15.0	1.80
Mineral Spirits	15.0	2.35
Texanol	10.0	1.26
PMA-30	.5	—
Ethylene Glycol	25.0	2.70
Hexylene Glycol	5.0	.65
Deefo 97-2	2.0	.14

	lb	gal
NH$_4$OH Conc.	2.0	.30
Titanox RANC	120.0	3.43
Busan 11-M 1	100.0	3.64
Barytes X-5-R	100.0	2.69
Mica 325 Mesh	25.0	1.06
Kadox #15	20.0	.42

Disperse on ball mill.

	lb	gal
Acronal 290 D	451.0	51.84
Morpholine ⎱ Premix	1.5	—
Oleic Acid ⎰	3.5	—
PVC (%)	30.2	
Nonvolatile (%)	53.8	
Nonvolatile by vol. (%)	37.8	
Wt./gal (lb)	11.3	

No. 2

(Red Oxide – Acrylic)

	lb	gal
Water	173.0	20.77
TKPP	.5	.06
8% Latekoll D	50.0	5.92
Mineral Spirits	15.0	2.35
NH$_4$OH Conc.	2.0	.30
25% Phtalopal SEB-M*	30.0	3.45
Texanol	10.0	1.26
Ethyl Glycol	25.0	2.70
Hexylene Glycol	5.0	.65
PMA-30	.5	—
Red Oxide - 3097	100.0	2.45
Oncor M-50	80.0	2.34
Barytes X-5-R	60.0	1.64
Mica 325 Mesh	25.0	1.06
Kadox #15	20.0	.42
Atomite	75.0	3.32

	lb	gal
Disperse on ball mill.		
Deefo 97-2	2.0	.14
Acronal 290 D	447.8	51.47
***25% Phtalopal SEB-M**		
Phtalopal SEB		250.0
Morpholine		70.0
Water		680.0
PVC (%)		30.8
Nonvolatile (%)		53.3
Nonvolatile by vol. (%)		36.5
Wt./gal (lb)		11.2

No. 3

(Chrome Yellow — Acrylic)

	lb	gal
Water	183.0	21.97
Latekoll D (8%)	40.0	4.73
Mineral Spirits	15.0	2.35
Ammonium Hydroxide (conc)	2.0	.30
Pigment Disperser A	15.0	1.80
Texanol	10.0	1.26
Ethylene Glycol	25.0	2.70
Hexylene Glycol	5.0	.65
PMA-30	.5	—
Chrome Yellow 45-090	100.0	2.27
Oncor M-50	80.0	2.34
Barytes X-5-R	60.0	1.64
Mica 325 Mesh	25.0	1.06
Kadox #15	20.0	.42
Atomite	75.0	3.32

Disperse on ball mill.

	lb	gal
Deefo 97-2	2.0	.14
Acronal 290 D	447.8	51.47
Morpholine	1.5	—
Oleic Acid	3.5	—
PVC (%)		30.1
Nonvolatile (%)		53.3
Nonvolatile by vol. (%)		37.3
Wt./gal (lb)		11.3

Acrylic Lacquer Aerosol Metal Coating

FORMULA No. 1

(Clear)

This is a low-cost system which provides antitarnish corrosion protection for metal surfaces.

		% in Aerosol
Concentrate:		55.0
Dow Corning 804 Resin	3.0%	
Acryloid B-66 Resin	4.3%	
Chlorothene Solvent	92.7%	
Propellent:		45.0
Isotron 12	100.0%	

Package:
Pressure fill in tin-plate cans with appropriate valve and button.

Procedure:
Dissolve **Acryloid B-66** resin in Chlorothene solvent. Mix in **Dow Corning 804 Resin**.

Directions for Use:
Shake well. Hold can 12 to 14 inches from surface to be sprayed.

Precautions:
Warning: Contents under pressure. Do not puncture. Exposure to heat

or prolonged exposure to sun may cause bursting. Do not throw into fire or incinerator. Keep from children.

	No. 2 *(Clear Flurocarbon)*		No. 3 *(Clear Hydrocarbon)*
Propellant:			
Lacquer Mix:			
Acryloid B-66 (40%)	50.0		43.75
Toluene	42.0		—
Xylene	—		42.25
Cellosolve Acetate	8.0		14.00
Spray Can (6 oz.) Charge:			
Lacquer	80.0 g		70.0 g
Freon 11	60.0 g		—
Freon 12	60.0 g		—
Methylene Chloride	—		21.0 g
Isobutane	—		42.0 g
Propane	—		7.0 g
Solids	8.0%		7.0%
Acryloid B-66	100%		100%
Volatiles	92.0%		93.0%
Toluene	31.3%	14.6%	
Xylene	—	23.7%	
Cellosolve Acetate	3.5%	7.9%	
Freon 11	32.6%	—	
Freon 12	32.6%	—	
Methylene Chloride	—	16.1%	
Isobutane	—	32.3%	
Propane	—	5.4%	

Aerosol Aluminum Enamel

(Acrylic)

Acrylic Resin (45% solids in VM&P naphtha)	225
Velsicol XL-30	—

VM&P Naphtha	465
Xylol	60
Toluene	—
Lactol Spirits	—
Chevron 298 E	—

Procedure:

Mix to clear solution. Slowly add to aluminum paste (below) with constant stirring.

Aerosol High-Temperature Aluminum Paint

(Acrylic)

		% in Aerosol
Concentrate:		55.0
Dow Corning 840 Resin (60%)	8.75%	
Aluminum Paste 1578 (75%)	8.75%	
Acryloid B-66 (40%)	14.90%	
Xylene	8.20%	
Toluene	59.40%	
Propellent:		45.0
Isotron 12	100.00%	

Package:

Package in tin-plate cans with appropriate valve and button. WARNING: Flammable.

Procedure:

Cold blend ingredients. Add agitation ball in each can.

Directions for Use:

Shake well. Hold can 12 to 14 in from surface to be sprayed.

Precautions:

Warning: Contents under pressure. Do not puncture. Exposure to heat or prolonged exposure to sun may cause bursting. Do not throw into fire or incinerator. Keep from children.

Lead Primer

FORMULA NO. 1

(Chlorinated Rubber)

Alloprene R20	11.0
Cereclor 42	4.4
Microtalc ATI	4.1
Barytes	19.6
Metallic Lead Paste (91% in **Cereclor**)	32.0
Thixcin E	0.36
Aromasol H	
White Spirit I	28.54

Thixcin E can be replaced by Thixcin R which is more commonly used in the paint industry.

NO. 2

(Chlorinated Rubber – Red)

Red Lead (97%)	698
M-P-A 60 (Xylene)	5
Dyphos	2
Copolymer 186	111
Parlon S-20 or **Alloprene 20** (30% N.V. in 80%	
Solvesso 100/20% Cellosolve)	198
80% Solvesso 150/20% Cellosolve	117 *
Epichlorhydrin	0.3
Lead Naphthenate (24%)	2.3
Cobalt Naphthenate (6%)	0.9
Manganese Naphthenate (6%)	0.6
Antiskinning Agent	2.7
Volatile and Driers (%)	23.4
Weight Ratio: Pigment/NV vehicle	81/19
PVC, Approximate (%)	35
Visc. (KU)	80

The above formula gives very fast dry coupled with good flexibility and excellent chemical resistance; grind temperature 130 F.

---

*For faster dry and/or spraying replace with **Solvesso 100**.

Lead Silico Chromate Primer

(Chlorinated Rubber)

Alloprene 20 cps	17.0
Cereclor 42	8.5
Thixatrol ST	0.6
Nonchalking Titanium Dioxide	5.0
Oncor M-50	16.3
Barytes	9.8
Propylene Oxide	0.1
Solvesso 100	32.2
Short Range Spirits	10.5
Wt. per gal (lb)	10.8
PVC (%)	30.5
Visc. (KU)	86
NV Wt. (%)	57.2
NV vol. (%)	34.2

This primer has been tested by brush and conventional air assisted spray application. It has not been tested for airless spray application.

Anticorrosive Primer

FORMULA No. 1

(Reddish Brown — Chlorinated Rubber)

Basic Lead Silico Chromate **Oncor M-50**	950
Red Iron Oxide (Siliceous, 85% FE_2O_3)	50
M-P-A 60 (Xylene)	15
Dyphos	6.1
Copolymer 186	252
Parlon S-20 or **Alloprene 20**) (30% NV in 80% **Solvesso 100**/20% **Cellosolve**)	452
Epichlorohydrin	0.5
Solvesso 150	263 *
Lead Naphthenate (24%)	3.1
Cobalt Naphthenate (6%)	1.3
Manganese Naphthenate (6%)	0.6
Antiskinning Agent	3.0
Volatile and Driers (%)	30.0
Weight Ratio (Pigment/NV Vehicle)	72.1/27.9

PVC, approximate (%)	38
Visc. (KU)	75

Very fast dry coupled with good flex, excellent chemical resistance; grind temperature 130 F.

The following pigment modifications are suggested to obtain alternate colors to the reddish brown submitted primers:

Orange: Remove iron oxide and use all **Oncor M-50** pigment.

Maroon: Reduce **Oncor M-50** to 650 parts and increase iron oxide to 350 parts (1000 parts total).

Gray: Reduce **Oncor M-50** to 684 parts, eliminate the iron oxide, and add 155 parts **Titanox RANC**, 51 parts zinc oxide, 98 parts magnesium silicate, 8 parts lampblack and 4 parts phthalocyanine blue.

------

*For faster dry and/or spraying replace with **Solvesso 100**.

No. 2

(Chlorinated Rubber)

	lb
Parlon S-20	50.14
Kippers Rezyl 896	99.73
Zinc Yellow	224.26
"1681" Red Oxide	45.62
3x **Asbestine**	179.63
Bentone 38	21.27
Amsco Solv-D	321.56
Solvesso 100	118.80
Turpentine	38.57
Epichlorhydrin	0.33
Dyphos	1.98

Anticorrosive Chlorinated Rubber Thick Coatings Primers

Formula	No. 1	No. 2	No. 3	No. 4
	(Red Lead)	*(Zinc Chromate)*	*(Metallic Lead)*	*(Zinc Rich)*
Alloprene 125 cps				2.3
Alloprene 40 cps				1.6
Alloprene 20 cps	14.5	17.8	15.2	
Cereclor 42	5.3	8.9	6.7	2.6
Red Lead	25.4			
Zinc Chromate		14.3		
Zinc Dust (Superfine)				74.8
Metallic Lead Paste			19.0	
(91% lead in **Cereclor 42**)				
Red Iron Oxide			4.8	
Bartyes	12.7	10.3	9.5	
C-3000 Mica			4.8	
Rutile Titanium Dioxide		5.2		
Epoxidized Soy Bean Oil			0.9	1.0
Thixatrol ST	0.5	0.8	0.5	0.2
Solvesso 100	31.2	32.2	29.0	17.5
Mineral Spirits	10.4	10.7	9.6	
Total lbs.	100.0	100.0	100.0	100.0
Wt./gal (lb)	12.1	10.9	11.9	22.7
Visc. (KU)	92	132	95	101
Nonvolatile weight (%)	58.4	57.1	61.4	82.5

Anticorrosive Chlorinated Rubber Topcoats

Formula	No. 1	No. 2	No. 3	No. 4
	Brush		*Airless Spray*	
	(White)	*(Red)*	*(Red)*	*(Gray)*
Alloprene 10 cps	19.4	19.4	13.6	12.0
Cereclor 70	12.9	12.9	7.9	7.0
Cereclor 42	6.5	6.5	4.5	4.0
Rutile Titanium Dioxide	15.2			14.0
Bartyes	11.5	16.1	15.8	14.0
Red Iron Oxide		10.8	15.8	

Anticorrosive Chlorinated Rubber Topcoats (cont'd)

	No. 1	No. 2	No. 3	No. 4
	Brush		*Airless Spray*	
	(White)	*(Red)*	*(Red)*	*(Gray)*
Carbon Black			0.4	0.2
Thixatrol ST	2.2	2.2	1.6	1.6
Xylene	32.3	32.1	32.3	37.8
Cellosolve Acetate			8.1	9.4
Wt./gal (lb)	11.6	11.8	11.7	10.8
Nonvolatile Wt. (%)	67.7	67.9	59.6	52.8

Thixotropic Zinc Chromate Primer

(Two Coat Maintenance Systems – Chlorinated Rubber)

Parlon S-10	28.68
Aroclor 1254	11.26
Aroclor 5460	7.08
Zinc Yellow **(Imperial Color No. X2127)**	12.52
Micro-Mica C-3000	13.43
Asbestine 325	12.52
Rutile Titanium Dioxide	4.90
Zinc Oxide (AZO-ZZZ-22)	6.35
Thixatrol ST	2.36
ERL-2774	0.72
Epichlorohydrin	0.18

Solvents

Toluene	100.0
Solids as prepared (Wt. %)	55.0
Solids as applied (Wt. %)	42.0
Solids as applied (vol. %)	23.0

Procedure:

Disperse the **Thixatrol ST** in one-half the solvent at a temperature of

115±5 F. Mix approximately 5 minutes on a Cowles Dissolver. Add **Parlon** and other vehicle solids. Maintain temperature at 115 F. Add dry pigment after **Parlon** and resins are dissolved. Continue high speed mixing until desired fineness of grind is obtained. Add remainder of solvent.

Chapter V

MASONRY PAINTS

A good masonry paint is specifically designed to handle the alkaline conditions in uncured cementitious substrates. The surface requires a high-binder primer because of its porous nature and resultant tendency to suck in liquids.

Exterior Flat Alkyd Primer for Masonry Surfaces

	lb	gal
Ti-Pure R-900	100	2.9
Drikalite	350	15.6
Water Ground Mica	50	2.1
Thixcin R	4	0.5
Metasol 57	4	0.2
Aroflat 3025 (40 N.V.)	450	63.0
Cobalt Naphthenate (6%)	1	0.1
Zirconium Naphthenate (6%)	4	0.5
Antiskinning Agent	1.5	0.2
Mineral Spirits	97	14.9
PVC (%)		49
Visc. (KU)		85

Exterior Flat Top Coat for Masonry Surfaces

(Alkyd)

	lb	gal
Ti-Pure R-900	200	5.8
Drikalite	375	16.7
Water Ground Mica	50	2.1
Thixcin R	4	0.5
Metasol 57	4	0.2
Aroflat 3025 (40 N.V.)	372	52.0
Cobalt Naphthenate (6%)	1	0.1
Zirconium Naphthenate (6%)	4	0.5
Antiskinning Agent	1.5	0.2
Mineral Spirits	143	21.9
PVC (%)		58
Visc. (KU)		85

Epoxy Coating

This is a general application spray coating for concrete blocks, appliance finishes, tank and drum linings, and protection or maintenance finishes for metal and wood. The **M-P-A** prevents pigment settling and sagging without undue increase in spraying viscosity.

A	**Versamid 140**	150
	Titanox RA-50	360
	Acicular ZnO (Eagle Picher #417)	90
	Xylene	57
	Butanol	45
	M-P-A (Xylene), or	9
	Thixatrol ST	5.4
B	**Gen Epoxy 185**	350
	Xylene	66

Note: In this formulation, **Thixatrol ST** is generally preferred for a ball mill grinding operation and M-P-A (Xylene) for a disperser type incorporation.

Concrete Tie-Coat

(Epoxy)

		gal
A	**ERL-2774**	437.44
	Silicone R-64	8.75
	Ti-Pure B-610	21.87
	Alpine Talc #111	43.74
	Thixatrol ST	17.50
B	**Thiokol LP-3**	437.44
	Alpine Talc #111	78.74
	DMP-30	43.74
	Carbon Black	1.44

Procedure:

Blend ingredients in A with 1/3 the resin and disperse to a 5H reading on the Hegman grind gage. Add remaining 2/3 of resin with good agitation. Repeat operation for B.

Application:

For adhesion of new concrete to old concrete. Spray.

Low-Temperature Masonry Sealant

(Epoxy)

A	**Resypox 1628**	100
	Yarmor 302 Pine Oil	12
	Surfex MM	28
	Silica 290	25
	Cab-O-Sil M-5	1.0
	Methanol	0.25
B	**Resycure 317**	31
	Yarmor 302	4
	Asbestine 3X	20
	Silica 290	25
	Cab-O-Sil M-5	2.5
	Methanol	1.0

Mixing Ratio:

2:1 by Wt.

Characteristics:

A Thixotropic filled paste which when cured yields a rigid masonry crack sealer capable of curing at temperatures down to 25°C.

Potlife: 45 g mass @ 75°F.:	15 min
45 g mass @ 35°F.:	24 hr
Shore 'D' Hardness after 24 hr @ 75°F.	88–90

Two-Component Masonry Coating

(Epoxy)

A	**Titanox Ranc**	100
	Lamp Black	2
	Nytal 300	575
	Resymide 1415	247
	High Flash Naphtha	75
	Cellosolve Solvent	90

Grind on three (3) — roll mill

B	Resypox 1302	225

Procedure:

Mix A and B and allow to stand for 30 min before application.

White Masonry Coating

(Epoxy)

Actual field tests have proved this to be a superior formulation for painting cinder and concrete block. High solids make one heavy coat enough for most applications. However, a grout or filler coat is recommended for very porous masonry surfaces.

Araldite 502	225
Rutile Titanium Dioxide	250
Bentone 27	7
Magnesium Silicate	212
Cellosolve	38
High-Flash Naphtha	31
CIBA Polyamide 815 X-70	214

Magnesium Silicate	212
Bentone 27	7
Cellosolve	58
High-Flash Naphtha	46

Primer for Concrete

(Epoxy)

		lb
A	**Epon 828**	92.0
	Epoxide 7	8.0
B	**Uni-Rez 2341-D**	50.0

Procedure:

Mix A and B together just prior to application. Apply to clean concrete using a trowel or squegee. Coverage is approximately 100 ft^2 per gal depending on the porosity of the concrete. The topping should be placed before the primer has cured.

Application:

Formulated to be applied by troweling to resurface eroded or abraded concrete floors. For application in chemical plants, meat processing and food packing plants where floors are subject to attack from exposure to various chemicals. Before application the flooring must be cleaned of all dirt and grease.

Masonry Patching Compound — Solventless and Sand Filled

(Epoxy)

		lb
A	**Epotuf 37-128**	220
	Thixatrol ST	8
B	**Epotuf Hardener 37-612**	132
	DMP-30	11
	Thixatrol ST	5.5
C	#1/4 Sand	1280

Procedure:

Disperse **Thixatrol ST** in **Epotuf 37-128** (A) on disperser type equipment. Repeat operation for B. The sand is added prior to use, at which time B is also added to A.

Application:

Masonry patching, highway repairs, bridge repairs, under block coatings. Trowel. The use of **Thixatrol ST** results in a pronounced improvement in the handling and workability of this compound.

White Masonry Surfacer

(Epoxy – Polyamide)

		lb/100 gal
A	**Titanox RA**	33.0
	Epotuf 37-128	417.0
	Epoxide 7	42.0
	Cab-O-Sil	24.0
	No. 219 Silica	334.0
	No. 325 Mesh Mica	50.0

Disperse on Cowles high speed disperser

Glycerin	1.5

Disperse additional 5 min

B	**Epotuf Hardener 37-612**	185.0
	Nytal 300	167.0
	Pennsalt S-2	10.0

Uses:

High quality surfacer for rough masonry substrates.

Floor Finish

FORMULA NO. 1

(Acrylic – Epoxy)

	lb
Titanium Dioxide (Rutile)[1]	100.0
Barium Sulfate[2]	20.0

	lb
Colloidal Silica[3]	10.0
Water	83.3
Thickener Sol'n. (2½%)[4]	83.3
Pigment Dispersant (35%)[5]	2.0
Pigment Suspension Agent[6]	3.0
Defoamer[7]	3.0
Preservative[8]	0.6

Grind — Pebble mill to fineness 7 + (N.S.):

	lb
Wetting Agent[9]	3.5
Acrylic Emulsion (46%)[10]	276.0
Defoamer[7]	3.0
Araldite PR-805 (50%)	254.0
Thickener Sol'n. (2½%)[4]	76.6
Cobalt (6%)[11]	2.0
Manganese Naphthenate (6%)	1.0
Ethylene Glycol	14.0
Tint with	
Black Aqueous Pigment	
Dispersion[12]	18.4

Performance:

Gloss, 60° Glossmeter	20
Drying Time: Set-to-Touch	8 min
Dry hard	3 hr
Dry through	6½ hr
Pencil Hardness, 7 days	B–HB

---

[1] Ti-Pure R-901
[2] Barytes No. 1
[3] Syloid 72
[4] Cellosize WP-4400
[5] Daxad 30 (35%)
[6] Emultex R
[7] AF-7
[8] PMA-18
[9] Triton X-100
[10] Rhoplex AC-61
[11] Cyclodex Cobalt
[12] Imperial X-2488

No. 2

(Epoxy – PVA)

Rutile Titanium Dioxide[1]	100.0
Barium Sulfate[2]	20.0
Colloidal Silica[3]	15.0
Water	83.3
Thickener Sol'n. (2½%)[4]	83.3
Pigment Dispersant (35%)[5]	2.0
Pigment Suspension Agent[6]	3.0
Ammonium Hydroxide (28%)	1.0
Defoamer[7]	3.0
Preservative[8]	0.6

Grind – Pebble Mill to Fineness 7 + (N.S.)

Ammonium Hydroxide (28%)	1.0
PVA Copolymer (55%)[9]	230.0
Araldite PR-805 (50%)	246.0
Wetting Agent[10]	2.0
Defoamer[7]	3.0
Thickener Sol'n. (2½%)[4]	75.5
Water	58.5
Cobalt (6%)[11]	2.0
Manganese Naphthenate (6%)	1.0
Ethylene Glycol	14.0
Tint with	
Black Aqueous Pigment	
Dispersion[12]	18.4

Performance:

Gloss, 60° Glossmeter		20
Drying Time:	Set-to-Touch	8 min
	Dry Hard	2 hr
	Dry Through	4½ hr
Pencil Hardness, 24 hr		2B-B

---

[1] Ti-Pure R-901	[7] AF-7
[2] Barytes No. 1	[8] PMA-18
[3] Syloid 72	[9] Everflex MF
[4] Cellosize WP-4400	[10] Stepanol B-153
[5] Daxad 30 (35%)	[11] Cyclodex Cobalt
[6] Emultex R	[12] Imperial X-2488

Epoxy Coating – Polyamide Cure

A	Versamid 140	150
	Titanox RA-50	360
	Acicular ZnO	90
	Xylene	57
	Butanol	45
	M-P-A 60 (Xylene) or	14.9
	Thixatrol ST	5.4
B	Gen Epoxy 185	350
	Xylene	66

In this formulation, **Thixatrol ST** is generally preferred for a ball mill grinding operation and **M-P-A 60** (Xylene) for a disperser-type incorporation.

Application:

This is a general application spray coating for concrete blocks, appliance finishes, tank and drum linings, and protection or maintenance finishes for metal and wood. The **M-P-A 60** prevents pigment settling and sagging without undue increase in spraying viscosity.

Clear High Solids Brushing Masonry Sealer

(Epoxy – Polyamide)

		lb/100 gal
A	Cab-O-Sil	28.0
	Epotuf 37-128	473.0
	No. 219 Silica Sand	372.0

Disperse on high speed Cowles disperser to 4–5 N.S. fineness

	Glycerine	1.9

Disperse additional 5 min

	MEK	74.0
B	Epotuf Hardener 37-612	126.0
	DMP-30	22.0
	Ethyl Alcohol	9.5

Uses:

Waterproofing sealant for cinderblock and other masonry surfaces.

Exterior Masonry Coating

Formula No. 1

(Latex)

	lb	gal
Ti-Pure R-900	200	5.9
Duramite	400	17.8
Thixcin R	5	0.6
Pliolite VT or **VT-AC**	90	10.5
Paroil 150	69	6.6
Mineral Spirits	372	58.6
Visc. (KU)	80–85	
PVC (%)	58.0	
Paint NV by vol. (%)	41.4	
Paint NV by Wt.	67.2	
Vehicle NV by Wt.	30.0	

No. 2

(Latex)

	lb	gal
Ti-Pure R-900	200	5.9
Duramite	400	17.8
Thixcin R	5	0.6
Pliolite S-5A or **AC**	91	10.5
Paroil 150	69	6.6
Mineral Spirits	201	30.8
High-Flash Naphtha	201	27.8
Visc. (KU)	80–85	
PVC (%)	58.0	
Paint NV by vol. (%)	41.4	
Paint NV by Wt. (%)	65.5	
Vehicle NV by Wt. (%)	28.5	

Stucco and Masonry Paint

FORMULA NO. 1

(Latex — Deep-Tone Green)

	lb/100 gal
Titanium Dioxide (Rutile NC)	120
Phthalocyanine Green Toner (**Du Pont GT-674-D**)	10
Lampblack (Superjet)	1
Calcium Carbonate (**Duramite**)	275
Mica (**Mineralite 3X**)	80
Diatomaceous Silica (**Celite 110**)	75
Thixcin R	5
Pliolite S-5A	86
Chlorinated Paraffin (40%)	43
Chlorinated Paraffin (70%)	43
High-Flash Naphtha	201
Mineral Spirits	201

PVC (%)	57.5
Solids (%)	64.7
Visc. (KU)	74

No. 2

(Latex — Deep-Tone Gray)

	lb/100 gal
Titanium Dioxide (Rutile NC)	120
Lampblack (Superjet)	8
Calcium Carbonate (**Duramite**)	275
Mica (**Mineralite 3X**)	80
Diatomaceous Silica (**Celite 110**)	75
Thixcin R	5
Pliolite S-5A	86
Chlorinated Paraffin (40%)	43
Chlorinated Paraffin (70%)	43
High-Flash Naphtha	201
Mineral Spirits	201

PVC (%)	57.5
Solids (%)	64.6
Visc. (KU)	74

No. 3

(Latex – Sun Yellow)

	lb/100 gal
Yellow Titanium Dioxide (Sun Yellow N)	142.0
Calcium Carbonate **(Duramite)**	249.0
Mica **(Mineralite 3X)**	72.0
Diatomaceous Silica **(Celite 110)**	68.0
Thixcin R	5.0
Pliolite S-5A	78.8
Chlorinated Paraffin (40%)	39.4
Chlorinated Paraffin (70%)	39.4
High-Flash Naphtha	214.0
Mineral Spirits	214.0

PVC (%)	57.4
Solids (%)	61.8
Visc. (%)	72

No. 4

(Latex – Deep-Tone Blue)

	lb/100 gal
Titanium Dioxide (Rutile NC)	153
Phthalacyanine Blue Toner **(Du Pont BT-284-D)**	3
Lampblack (Superjet)	0.4
Calcium Carbonate **(Duramite)**	260
Mica **(Mineralite 3X)**	75
Diatomaceous Silica **(Celite 110)**	70
Thixcin R	5
Pliolite S-5A	90
Chlorinated Paraffin (40%)	45
Chlorinated Paraffin (70%)	45
High-Flash Naphtha	200
Mineral Spirits	200

PVC (%)	56
Solids (%)	65.1
Visc. (KU)	75

No. 5

(Latex – Deep-Tone Red)

	lb/100 gal
Red Iron Oxide **(Mapico 297)**	125
Calcium Carbonate **(Duramite)**	275
Mica **(Mineralite 3X)**	75
Diatomaceous Silica **(Celite 110)**	75
Thixcin R	5
Pliolite S-5A	81
Chlorinated Paraffin (40%)	40.5
Chlorinated Paraffin (70%)	40.5
Soya Lecithin	10
High-Flash Naphtha	210
Mineral Spirits	210

PVC (%)	57.6
Solids (%)	63.4
Visc. (%)	74

Exterior Latex Masonry Paint Tint

FORMULA	No. 1	No. 2
	(Blue)	*(Pink)*
	lb/100 gal	
Titanium Dioxide (Rutile NC)	141.5	116.5
Phthalo Blue	1.8	—
Calcium Carbonate (low oil absorption)	236.0	261.0
Red Iron Oxide	—	1.7
Mica **(Mineralite 3X)**	68.0	71.0
Celite 110	63.5	71.0
Soya Lecithin*	4.0	5.0
Thixcin R	4.6	4.7
Pliolite AC	94.0	94.0
Chlorinated Paraffin (40%)	47.0	47.0
Chlorinated Paraffin (70%)	47.0	47.0
Solvesso 100	203.0	199.0
Mineral Spirits	203.0	199.0

	No. 1	No. 2
	(Blue)	*(Pink)*
	lb/100 gal	
PVC (%)	52	53
Solids (Wt. %)	63.5	64.4

---

*Variable dependent upon color system used 3–10 lb soya lecithin per 200 gal is recommended range of use.

Note: Above formulation is not recommended for use as pastel yellow.

No. 3

	lb/100 gal
Rutile Titanium Dioxide	115
Calcium Carbonate	312
Mica	72
Celite 110	72
Soya Lecithin	*
Thixcin	5
Pliolite S-5 or **Pliolite AC**	82
Chlorinated Paraffin (40%)	41
Chlorinated Paraffin (70%)	41
High-Flash Naphtha	202
Mineral Spirits	202

---

*Variable depending upon color

Note: For optimum holdout and suitable viscosity control the oil absorption of the pigment is critical. The calcium carbonate should have an oil absorption of 5–6, mica should be 31, and diatomaceous silica that of 170.

Concrete Floor Latex Enamels

	FORMULA	No. 1	No. 2	No. 3
		(Dark Red)	(Tile Red)	(Oak)
Add ingredients in order shown:				
Aromatic High-Flash Naphtha		200.0	200.0	200.0
Marbon 9200 MV		150.0	150.0	150.0
Mix to clear solution:				
Chlorinated Paraffin (50%)		75.0	75.0	75.0
Raw Linseed Oil		15.0	15.0	15.0
Zinc Oxide		20.0	20.0	20.0
Aluminum Silicate		65.0	65.0	40.0
Red Iron Oxide		125.0	—	—
Red Iron Oxide		—	125.0	—
Yellow Iron Oxide		—	—	150.0
*Suspension Additive		1.0	1.0	1.0
*Denatured Alcohol		1.0	1.0	1.0
Lecithin		2.0	2.0	2.0

(*Prewet)

Mix to smooth paste and grind on 3-roll mill to fineness of 7 N.S.

	No. 1	No. 2	No. 3
Mineral Spirits	230.0	230.0	230.0
Aromatic High-Flash Naphtha	30.0	30.0	30.0
Cobalt Tallate (6%)	4.0	4.0	4.0
Total Solids (%)	49.8	49.8	49.8
PVC (%)	19.8	20.4	19.6
Wt. per gal (lb)	9.1	9.1	9.1
Visc. (KU)	73	71	72
Gloss 60°	82	87	81
Drying Time:			
Set-to-Touch	40 min	40 min	40 min
Tack Free	2.5 hr	2.5 hr	2.5 hr
Dry Hard	4.0 hr	4.0 hr	4.0 hr
Dry Through	4.0 hr	4.0 hr	4.0 hr

	No. 4 (Medium Grey)	No. 5 (Light Grey)	No. 6 (Dark Green)
Aromatic High-Flash Naphtha	200.0	200.0	200.0
Marbon 9200 MV	150.0	150.0	150.0
Mix to clear sol'n.:			
Chlorinated Paraffin (50%)	75.0	75.0	75.0
Raw Linseed Oil	15.0	15.0	15.0
Zinc Oxide	20.0	20.0	20.0
Aluminum Silicate	65.0	40.0	65.0
Titanium Dioxide (rutile)	125.0	175.0	–
Chrome Green (medium)	–	–	100.0
*Suspension Additive	1.0	1.0	1.0
*Denatured Alcohol	1.0	1.0	1.0
Antifloating Additive	1.0	1.0	–
Lecithin	1.0	1.0	2.0

(*Prewet)

Mix to smooth paste and grind on 3-roll mill to fineness of 7 N.S.

Mineral Spirits	230.0	230.0	230.0
Aromatic High-Flash Naphtha	30.0	30.0	30.0
Cobalt Tallate (6%)	4.0	4.0	4.0
Lampblack in Oil	15.0	7.0	–
Red Iron Oxide in Oil	2.0	–	–
Total Solids (%)	50.3	51.1	47.9
PVC (%)	21.1	22.6	20.0
Wt./gal (lb)	9.1	9.2	8.9
Visc. (KU)	71	72	73
Gloss 60°	90	92	81
Drying Time:			
Set-to-Touch	40 min	40 min	40 min
Tack Free	2.5 hr	2.5 hr	2.5 hr
Dry Hard	4.0 hr	4.0 hr	4.0 hr
Dry Through	4.0 hr	4.0 hr	4.0 hr

Exterior Latex Masonry Paint

FORMULA No. 1

(White)

	lb/100 gal
Pliolite S-5 or **Pliolite AC**	73.5
Rutile Titanium Dioxide	177.4
Zinc Oxide	50.5
Asbestine	204.1
Celite 110	75.9
Chlorinated Paraffin (40%)	35.3
Chlorinated Paraffin (70%)	35.3
Solvesso 100	225.9
Mineral Spirits #10	225.9

Note: **Thixcin** at 4–5 lb per 100 gal or 20 lb of a 10% **Pliolite S-3** gel may be used to adjust viscosity. If **Pliolite S-3** gel is used, hold out 9 lb of each solvent.

Pliolite S-3 Gel	
Pliolite	10
MS #10	45
Solvesso 100	45

Add **Pliolite S-3** to **MS #10** under agitation then add **Solvesso 100**

Note: Not suitable for use as a tint base.

No. 2

(Deep-Tone Blue)

Rutile Titanium Dioxide	153.0
Phthalo Blue Toner	3.0
Lampblack	0.4
Calcium Carbonate	260.0
Mica **(Mineralite 3x)**	75.0
Celite 110	70.0
Thixcin R	5.0
Pliolite AC	100.0
Chlorinated Paraffin (40%)	50.0

Chlorinated Paraffin (70%)	50.0
Soya Lecithin	2.0
High-Flash Naphtha	193.0
Mineral Spirits	193.0

No. 3

(Deep-Tone Brown)

Brown Iron Oxide	115
Calcium Carbonate	288
Mica (**Mineralite 3x**)	72
Celite 110	72
Thixcin R	5
Pliolite AC	98
Chlorinated Paraffin (40%)	49
Chlorinated Paraffin (70%)	49
Lecithin	10
High-Flash Naphtha	194
Mineral Spirits	194

	No. 4	No. 5	No. 6	No. 7
	(Green)	*(Flamingo)*	*(Aqua)*	*(Yellow)*

Add ingredients in order shown:

Mineral Spirits	195.0	195.0	195.0	195.0
Marbon 1100T MV	83.5	83.5	83.5	83.5

Mix to clear sol'n.

Chlorinated Paraffin (50%)	83.5	83.5	83.5	83.5
Titanium Dioxide (rutile nonchalking)	160.0	160.0	160.0	160.0
Calcium Carbonate	197.0	197.0	197.0	197.0
Zinc Oxide	50.0	50.0	50.0	50.0
Mica	75.0	75.0	75.0	75.0
Diatomaceous Silica	75.0	75.0	75.0	75.0
Suspension Additive	4.0	4.0	4.0	4.0
Lecithin	4.0	4.0	4.0	4.0
Denatured Alcohol	2.0	2.0	2.0	2.0

(*Prewet)

	No. 4	No. 5	No. 6	No. 7
	(Green)	*(Flamingo)*	*(Aqua)*	*(Yellow)*

Mix to smooth paste and grind on 3-roll mill to fineness of 3–4 N.S.

	No. 4	No. 5	No. 6	No. 7
Mineral Spirits	195.0	195.0	195.0	185.0
Phthalocyanine Green in Oil	5.0	–	3.0	–
Yellow Iron Oxide in Oil	9.0	6.0	5.0	3.0
Lampblack in Oil	1.0	–	–	–
Red Iron Oxide in Oil	–	5.0	–	0.5
Phthalocyanine Blue in Oil	–	–	2.0	–
Hansa Yellow in Oil	–	–	–	15.0
Total Solids (%)	65.6	65.4	65.3	66.3
PVC (%)	51.9	52.3	52.3	52.0
Wt./gal (lb)	11.3	11.3	11.3	11.3
Visc. (KU)	78	78	78	78
Gloss 60°	2	2	2	2
Sheen 58°	3	3	3	3
Drying time:				
Set-to-Touch	20 min	20 min	20 min	20 min
Tack Free	30 min	30 min	30 min	30 min
Dry Hard	35 min	35 min	35 min	35 min
Dry Through	75 min	75 min	75 min	75 min

Latex Enamels

Formula No. 1

(Deep-Tone Gray)

	lb/100 gal
Titanium Dioxide (Rutile NC)	120
Lampblack (Superjet)	8
Calcium Carbonate (**Duramite**)	275
Mica (**Mineralite 3X**)	80
Diatomaceous Silica (**Celite 110**)	75
Thixcin R	5

	lb/100 gal
Pliolite S-5A	86
Chlorinated Paraffin (40%)	43
Chlorinated Paraffin (70%)	43
Soya Lecithin*	2
High-Flash Naphtha	201
Mineral Spirits	201
PVC (%)	57.5
Solids (%)	64.6
Visc. (KU)	74

*Soya lecithin improves the color uniformity of these inorganic pigment systems. For best results the lecithin should be added to the pigment grind.

No. 2

(Deep-Tone Blue)

	lb/100 gal
Titanium Dioxide (Rutile NC)	153
Phthalocyanine Blue Toner (**DuPont BT-284-D**)	3
Lampblack (Superjet)	0.4
Calcium Carbonate (**Duramite**)	260
Mica (**Mineralite 3X**)	75
Diatomaceous Silica (**Celite 110**)	70
Thixcin R	5
Pliolite S-5A	90
Chlorinated Paraffin (40%)	45
Chlorinated Paraffin (70%)	45
Soya Lecithin	2
High-Flash Naphtha	200
Mineral Spirits	200
PVC (%)	56
Solids (%)	65.1
Visc. (KU)	75

No. 3

(Green Gold)

	lb/100 gal
Titanium Dioxide (Rutile NC)	150.0
Metalized Azo Yellow **(DuPont YT-562D)**	7.5
Calcium Carbonate **(Duramite)**	250.0
Mica **(Mineralite 3X)**	70.0
Diatomaceous Silica **(Celite 110)**	70.0
Thixcin R	5.0
Pliolite S-5A	82.0
Chlorinated Paraffin (40%)	41.0
Chlorinated Paraffin (70%)	41.0
Soya Lecithin	2.0
High-Flash Naphtha	212.0
Mineral Spirits	212.0
PVC (%)	57.7
Solids (%)	62.8
Visc. (KU)	76

No. 4

(Hansa Yellow)

	lb/100 gal
Titanium Dioxide (Rutile NC)	113.0
Hansa Yellow G **(DuPont YT-445D)**	28.0
Calcium Carbonate **(Duramite)**	236.0
Mica **(Mineralite 3X)**	66.0
Diatomaceous Silica **(Celite 110)**	66.0
Thixcin R	5.0
Pliolite S-5A	81.2
Chlorinated Paraffin (40%)	40.6
Chlorinated Paraffin (70%)	40.6
Soya Lecithin	2.0
High-Flash Naphtha	208.0
Mineral Spirits	208.0
PVC (%)	58.0
Solids (%)	61.9
Visc. (KU)	78

No. 5

(Sun Yellow)

	lb/100 gal
Yellow Titanium Dioxide (Sun Yellow N)	142.0
Calcium Carbonate (Duramite)	249.0
Mica (Mineralite 3X)	72.0
Diatomaceous Silica (Celite 110)	68.0
Thixcin R	5.0
Pliolite S-5A	78.8
Chlorinated Paraffin (40%)	39.4
Chlorinated Paraffin (70%)	39.4
Soya Lecithin	2.0
High-Flash Naphtha	214.0
Mineral Spirits	214.0
PVC (%)	57.4
Solids (%)	61.8
Visc. (KU)	72

No. 6

(Chalk-Resistant White)

	lb/100 gal
Titanium Dioxide (Rutile NC)	175
Zinc Oxide	50
Talc (Asbestine 3X)	205
Diatomaceous Silica (Celite 110)	75
Thixcin R*	5
Pliolite S-5	75
Chlorinated Paraffin (40%)	35
Chlorinated Paraffin (70%)	35
High-Flash Naphtha	228
Mineral Spirits	228
PVC (%)	56.5
Solids (%)	60.0
Visc. (KU)	70

*For Pebble mill grind; for Cowles Dissolver, use **Thixatrol ST** or **Bentone 38.**

Concrete Floor Enamels

(Chlorinated Rubber)

FORMULA	No. 1	No. 2	No. 3
Ti-Pure R-900	175.0	175.0	175.0
Kadox 515 Zinc Oxide	20.0	20.0	20.0
Chlorinated Rubber (20 cps)	130.0	80.0	40.0
Chlorinated Rubber (125 cps)	50.0	80.0	100.0
Velsicol XL-30	20.0	40.0	60.0
Chlorinated Biphenyl	125.0	125.0	140.0
Chlorinated Triphenyl	54.0	54.0	39.0
Solvesso 100	368.0	368.0	368.0
Mineral Spirits	57.0	57.0	57.0
Epichlorohydrin	1.0	1.0	1.0
Total Weight (lb)	1000.0	1000.0	1000.0
Total Yield (gal)	96.1	96.7	97.6
Total Solids (%)	57.4	57.4	57.4

Concrete Curing Compounds

Concrete curing compounds are essentially very low cost coatings which can be sprayed on wet concrete surfaces immediately after completion of the finishing operations on slab concrete, or on formed concrete immediately after removal of the forms. The major function of these compounds is to form a quick impervious film which prevents any substantial loss of water by evaporation from the concrete. In this way there will be sufficient water present in the mass to allow an uninterrupted hydration of the cement during the curing period or until satisfactory strength is developed throughout the concrete mass.

Since economy is an essential characteristic of these compounds it is desirable to pigment the binder to as great an extent as possible with mostly low cost easy to wet extender pigments without creating porosity in the resulting dried film. By thus pigmenting the resinous binders used in these compounds it is possible to 1) have very low raw material costs per gallon, 2) have higher film solids per gallon, and 3) increase the rate of dry through reducing the amount of solvent present per gallon. These advantages can be attained without affecting the other properties providing the critical pigment volume concentration is not exceeded.

Concrete curing compounds are made as clear, white, or cured concrete colored coatings. The advantage of the white or light colored compounds is that they reduce the temperature increase of concrete slabs when exposed to the sun, thereby minimizing the expansion and contraction stresses resulting from extreme temperature changes. The white or colored compounds are obviously higher in raw material cost than those which do not contain prime pigments.

	Formula No. 1 (White)		No. 2 (Translucent)	
	lb	gal	lb	gal
Piccopale 100 (70% N.V.)	159	21.0	159	21.0
Unitol-R	41	5.0	41	5.0
TiPure R-900	50	1.4	–	–
Drikalite	750	33.4	800	35.6
Thixcin-R	4	0.5	4	0.5
Mineral Spirits	252	38.7	247	37.9

	No. 3 (Hydrocarbon – White)		No. 4 (Hydrocarbon – Translucent)	
	lb	gal	lb	gal
Piccopale 100 (70% NV)	159	21.0	159	21.0
Unitol-R	41	5.0	41	5.0
Ti-Pure R-900	50	1.4	–	–
Drikalite	750	33.4	800	35.6
Thixcin R	4	0.5	4	0.5
Mineral Spirits	252	38.7	247	37.9

White Masonry Paint – Light Tint Base

(Vinyl Acrylic)

	lb	gal
Water	125	15.0
KTPP	1	.1
Tamol 731-25	9	1.0

	lb	gal
Triton X-102	4	.5
Propylene Glycol	26	3.0
Foamicide **Nopco** 1419A	2	.3
Ti-Pure R-901	275	8.25
Whiting **Snowflake**	50	2.2
Micro **Mica C-1000**	25	1.1
Silica Gold Bond R	25	1.1
2.5% **Cellosize QP** (15,000 Sol'n.)	100	12.0
Mildewcide Metasol 57 Sol'n.	4	.5
Water	50	6.0
2.5 **Cellosize QP** (15,000 Sol'n.)	75	9.0
Texanol	20	2.5
Varaqua 936	382	42.0

PVC (%)	36.2
Total N.V. by Wt. (%)	50.8
Wt. per gal (lb)	11.20
Visc. (KU)	85–90
Brushing	Excellent
Leveling	Very Good
Sagging Resistance	Excellent
Settling	None
Skinning	None
Color Uniformity	Excellent
Recoat Uniformity	Excellent
Sheen	Good
Adhesion on Enameled Wood	Excellent

Prototypes of this paint have excellent film integrity when exposed for four years at 45° South in Los Angeles.

Concrete Floor Paints

Floor paints in general are characterized by their ability to have 1) good water resistance, 2) good adhesion to the applied surfaces, 3) good leveling and flow (whether brushed or rolled), 4) reasonable amount of gloss, 5) good scrubbability and 6) good mar and general wear resistance.

The use of a fine particle sized ground calcite in combination with the proper level of color pigment and a nonchalking grade of rutile pigment (i.e. if used), formulated at about 30% pigment volume concentration produces paints possessing the above mentioned characteristics.

Concrete Floor Paint

FORMULA	No. 1 (Gray)		No. 2 (Red)		No. 3 (Green)	
	lb	gal	lb	gal	lb	gal
TiPure R-900	100	2.9	—	—	—	—
Atomite	200	8.9	200	8.9	200	8.9
Daxad 30	5	0.5	5	0.5	5	0.5
Ethylene Glycol	28	3.0	28	3.0	28	3.0
Colloid 606	2	0.3	2	0.3	2	0.3
Igepal CO-610	2	0.2	2	0.2	2	0.3
Butyl Cellosolve Acetate	24	3.0	24	3.0	24	3.0
Troysan PMA-30	.3	—	.3	—	.3	—
Natrosol 250 H.R. (3%)	33	4.0	33	4.0	33	4.0
Lamp Black dispersion	13	1.6	—	—	—	—
Red Iron Oxide RCI 1380	—	—	75	1.9	—	—
Green Chrome Oxide X-1134	—	—	—	—	85	2.0
Rhoplex AC-61 (46% N.V.)	551	62.0	507	57.0	551	62.0
Water	113	13.6	177	21.2	134	16.1
PVC (%)	31.1		30.9		29.5	
Visc. (KU)	66		66		61	

White Masonry Paint

(PV Acetate)

	lb	gal
Premix:		
Water	262.0	31.50
Strodex PK-90 Dispersant	2.0	0.21
Ammonia (28%)	4.0	0.50
Triton X-100 Surfactant	5.0	0.60
Potassium Tripolyphosphate	1.0	0.05
Increase speed and add slowly:		
Chemacoil TA-100	78.0	10.00
Disperse pigments:		
Ti-Pure R-901 (titanium dioxide)	200.0	5.72
Nytal 300 Talc	200.0	8.40
Duramite Dry Ground Whiting	150.0	6.65
Slow mixing speed and add slowly:		
Colloid 581-B Defoamer	1.0	0.13
Ethylene Glycol	18.6	2.00
Nuodex PMA-18 Preservative	1.5	0.15
Cellosize QP-15000 (2%) Thickener	126.0	15.00
Everflex BG Vinyl Acetate		
Copolymer Emulsion (52%)	188.0	20,80
Super-Cobalt Drier	0.5	0.06

Visc. (KU) — Fresh	87
Overnight	90
4 days	90
PVC (%)	51.0

White Primer for Cement

(PVC)

	lb	gal
Vinoflex MP 400 (Medium)	124.0	12.00
Ketone Resin N	41.0	4.21

	lb	gal
Lutonal M-40 (50% Xylene)	82.0	10.35
Xylene	82.0	11.34
Butyl Acetate	82.0	11.25
Cellosolve Acetate	41.0	5.30
Titanox CLNC	150.0	4.39
Nytal 300	115.0	4.84
Dicalite WB-5	66.0	3.44
Bentone 38	1.5	.10

Procedure:
Ground 16 hr in a high density alumina jar mill

Then add:

Mineral Spirits	165.0	25.22
PVC (%)	31.0	
Nonvolatile (%)	58.3	
Nonvolatile by vol. (%)	44.5	
Wt./gal (lb)	10.27	
Visc. (KU)	84	

Chapter VI

WOOD COATINGS

Because of its hygroscopic nature, wood is dimensionally unstable. Thus a wood coating must be elastic to avoid cracking and it must adhere to prevent flaking and peeling. In order to prevent blistering, wood paint should be able to have water vapor pass through it in handling normal atmospheric humidity.

Mahogany Wood Filler

(Oil)

Silica	650.0
Acicular Talc (X fineness)	150.0
Calcium Linoleate Pulp (70% water)	20.0
Raw Linseed Oil	77.6
Gloss Oil Varnish (60% solids)	121.0
Mineral Spirits	64.7
Burnt Umber Paste in Oil	23.1
Maroon Toner Paste in Oil	1.9
Carbon Black Paste in Oil	7.0
Brown Toner Paste in Oil (Bleeding Type)	5.0

Burnt umber has drying action, eliminating the need for added drier.

Brown Toner consists of:

Diethylene Glycol Monoethyl Ether	6
Ethylene Glycol	6

194

| Methanol | 78 |
| Toluol | 10 |

Procedure:

Melt all colors and powders together in a pebble mill. Then mix the liquids and the oils together — if necessary — heat the mixture until dissolved. Mix with the ground colors and powders.

Blister-Resistant Oil-Based Primer

It has been well established that it is best to use a special exterior house paint blister resistant primer on wood substrates 1) that are unpainted, 2) that have had their previous coats removed for repainting (i.e., burned off), and 3) that have as a general condition erosion to the substrate. This is particularly important where the conventional types of latex emulsion paints and the ethylene-vinyl-acetate–based paints systems are to be used as top coats. It is, in fact, suggested that heavily chalked surfaces (on wood substrates) be first coated with such a primer in order to insure proper adhesion of these exterior latex emulsion top coats. The important characteristics that primers should have are 1) adhesion, 2) flexibility, 3) penetration control, 4) chalk resistance, 5) a type of surface to which top coats can adhere well, 6) blister and crack resistance, and 7) good application properties, thus careful selection of ingredients is essential.

Pigmentation: Ground calcium carbonate (e.g., **Duramite**) in combination with a maximum chalk resistant grade of rutile titanium dioxide and lead pigments (e.g., basic carbonate of white lead) have proven to be an excellent choice of pigments in such primers. These pigments are easily wetted by the binder and hence offer the desired penetration control with maximum chalk resistance. **Duramite** aids in forming the type of surface in primers to which a top coat will adhere easily. Lead pigments, and to a lesser degree, ground calcite, have low water absorption properties and hence develop better blister and crack resistance. Furthermore, lead pigments contribute to more film flexibility and help to control mildew growth as do ground calcite pigments. The presence of any zinc oxide is not recommended in blister resistant primers, since it tends to destroy blister and crack resistant properties. However with the absence of zinc oxide in such primers it is suggested that organic mildewicide be employed as required to effectively maintain mildew resistance.

Binders: A specific blend of refined or raw linseed oil with heat bodied linseed oil should be carefully chosen such that there is a limited penetration of the binder into the wood assuring good adhesion to this substrate and yet maintaining sufficient holdout to establish a good foundation for the finish coat. The choice of oil blend is also important so that the best possible leveling properties are obtained.

Oil-Based Primer

	lb	gal
TiPure R-900	150	4.4
Basic Carbonate White Lead	200	3.7
Duramite	600	26.7
Aluminum Stearate G. M.	2	0.2
Refined Linseed Oil	247	32.0
Z-3 Bodied Linseed Oil	128	16.0
Lead Naphthenate (24%)	8	0.8
Manganese Naphthenate (6%)	1	0.1
Metasol 57	4	0.2
Antiskinning Agent	1	0.1
Mineral Spirits	103	15.9
PVC (%)		41.9
Visc. (KU)		85

Redwood Finish

(Alkyd)

	lb	gal
Fales Thickener	6	
Ven. Red	36	
Yellow Oxide	4	
Chlorox 70	20	
Gm Al Stearate	4	
Yelkin	4	
FAGL 60 (Long Oil Alkyd)		31
Mineral Spirits		62
Celite 281	35	
Raw Linseed Oil		5
Cobalt (6%)	5	
Lead (25%)	6	
P.M.O. 30	2	
Snowflake	50	
A.S.A.		

Exterior Wood Stain

(Alkyd)

	lb
Chromium Oxide	50
Alkyd Vehicle (70% N.V.)	350
Mineral Spirits	280
Amsco D	100
Pentachlorophenol Solution (5%)	5
Post-4	8
Zinc Naphthenate (24%)	5.2
Cobalt Naphthenate (6%)	1.8

Interior Fast-Dry Lumber Primer

(Alkyd)

	lb	gal
Rutile TiO_2	200	5.72
Gamaco	450	20.03
Bentone 38	4	.30
Varkyd 586-50X	472	57.50
SoCal 2	147	20.50
Lead Naphthenate (24%)	5.00	.60
Cobalt Naphthenate (6%)	2.25	.10
Exkin #2	1.00	—

PVC (%)	51
Vehicle NV (% Wt.)	18.5
Total N.V. (Wt.)	70
Wt. per gal (lb)	12.5
Visc. (KU)	80–90
Reduction	175# **SoCal** 2 to 100 gal base
Reduction Visc.	20 s #4 Ford Cup

Molding Primer

(Alkyd)

	lb	gal
Varkyd 586-50X	266.0	32.40
SoCal #1	116.0	16.25

	lb	gal
Lead (24%)	2.5	23 fl. oz.
Soya Lecithin	8.0	1.0
Meta 571	8.0	1.0
R-760	95.0	2.72
RCHT	778.0	28.71
Desertalc 706	150.0	6.50

Grind to 6 min

	lb	gal
Varkyd 586-50X	181.0	22.10
SoCal #1	206.0	29.50
Cobalt (6%)	0.5	5 fl. oz.

PVC (%)	60.5
Vechicle N.V. (% Wt.)	30.2
Total NV (% Wt.)	69.8
Wt. per gal (lb)	12.9
Visc. (KU)	70
Gloss	Flat
Grind	6 min

Primer for Wood or Composition Board

(Alkyd)

	lb	gal
Ti-Pure R-960 Titanium Dioxide	216.0	6.57
Barytes X5R	108.0	3.01
Talc	54.0	2.32
Duramac 2482 Alkyd Resin (50%)	298.0	36.12
Plaskon 3353 Urea Coating Resin (50%)	81.0	9.62
Bakelite VAGH Vinyl Resin	65.0	5.61
Butanol	86.0	12.84
1-Nitropropane	43.0	5.15
Acetone	59.0	8.90
Butyl Acetate	59.0	8.08
Adamac Catalyst 20	13.0	1.78

Wt. per gal (lb)	10.82
Total solids (% Wt.)	54.71
% by Vol.	36.92

Baking schedule to F pencil hardness:

With **Catalyst 20**	4 min at 300 F
Without Catalyst	20 min at 300 F

Properties on **Bonderite** steel panel baked 30 min at 140 F

Film Thickness	0.6 mil
Pencil Hardness	2H
60° Gloss Reading	21
Impact, Direct	16 in.–lb
Reverse	<4 in.–lb
Adhesion	100%
Flexibility on Mandrel	OK

Unpigmented Coating for Wood and Metal

(Epoxy)

A	**Araldite 571 CX-80**	369.0
	Xylene	81.0
	Diacetone Alcohol	40.5
	Flow Control Agent	14.7
B	**Araldite Hardener 820**	246.0
	Xylene	54.0
	n-Butanol	27.0

Unpigmented Finish for Wood

(Epoxy)

A	**Araldite 7072**	262.0
	Xylene	57.0
	Methyl Cellosolve	26.0
	DIBK	23.0
	MIBK	57.0
	Flow Control Agent	5.0
B	**Araldite Hardener 820**	176.0
	Methyl Cellosolve	65.0
	Xylene	74.0
	DIBK	63.0

Clear Chemical-Resistant Coating for Wood and Metal

(Epoxy)

	lb
Epoxy Resin	100.00
Beetle (urea-formaldehyde resin)	5.27
Toluene	41.80
Methyl Isobutyl Ketone	19.80
Butanol	19.65
Butyl **Cellosolve** – Glycol Ether	2.26

Add just prior to use:

	lb
Diethylene Triamine	6.00
Butanol	2.04
Toluene	2.04
Visc. (cps)	55–65
Nonvolatile (%)	55
Specific Gravity	1.00
Color (Gardner)	3–5

White Maintenance Paint

(Epoxy)

		lb	gal
A	Rutile Titanium Dioxide	331.47	9.46
	Epoxy Resin	426.44	46.77
	Beetle	16.11	1.89
	Methyl Ethyl Ketone	95.72	14.26
	Cellosolve Solvent	47.39	6.11
	Butyl **Cellosolve**	11.36	1.51
B	**Polyamide 325**	106.14	13.11
	Xylene	26.82	3.72
	Cellosolve Solvent	24.49	3.17

Procedure:

Pebble mill grind A until a good dispersion is obtained. The hardener portion, B, can be mixed on any standard mixing equipment. Do not mix the two components until ready to apply.

Mixing Ratio:	By Wt.	By Vol.
Resin Portion	5.9 lb	0.8 gal
Hardener Portion	1.0 lb	0.2 gal

After the two components are mixed, a 30 min induction period is recommended before application of the coating.

Polyurethane Wood Finish

	FORMULA No. 1	No. 2
Polyol Component:		
Desmophen 650A	26.0	—
Epoxy Resin 1009	—	25.0
EAB 381-2 (10% in ethyl acetate)	1.0	—
Baysilone OF/OR311 (1% in ethyl acetate)	—	0.5
Modaflow (5% in ethyl acetate)	0.1	0.1
Solvent Blend	39.9*	46.4**
Desmorapid PP (10% in ethyl acetate)	2.5	2.0
Isocyanate Component:		
Desmodur N	30.5	26.0
NCO/OH	1.0	1.0
Solids % ca	49	45

Formula No. 1 — used where a light-stable coating with highest degree of resistance to weathering, chemicals and solvents is required.

No. 2 — used where good light stability and a high resistance to chemicals and solvents, especially alkalies, is required.

Note: Reduce formulations to 30–35% solids for spray application.

*Solvent Blend — ethylglycol acetate/xylol/MIBK 1 : 1 : 1
**Solvent Blend — ethylglycol acetate/xylol 2 : 1

	No. 3	No. 4	No. 5
Polyol Component:			
Desmophen 1100	23.5	–	–
E-380	–	42.5	33.0
EAB 381-2 (10% in ethyl acetate)	4.0	4.0	3.5
Modaflow (5% in ethyl acetate)	0.5	0.5	0.5
Solvent Blend	37.0*	19.0*	26.0**
Isocyanate Component:			
Mondur HC	36.0	34.0	–
Desmodur IL	–	–	37.0
NCO/OH	1.1	1.1	1.1
Solids % ca	45	52	43

Formula No. 3 — for use where fast-curing wood finishes of good hardness and chemical resistance are required.

No. 4 — for use where fast-curing wood finishes of moderate chemical resistance and good hardness are required.

No. 5 — for use where extremely fast-curing wood finishes are required, along with good hardness properties.

Note: Reduce formulations No. 3 and No. 4 to about 35% solids for spray application. Reduce Formula No. 5 to 30% solids.

*Solvent blend — ethylglycol acetate/xylol 2:1
**Solvent blend — ethylglycol acetate/MIBK 2:1

No. 6

Multron R-12A	275
Ethylglycol Acetate	175
Toluol	179
Modaflow	1
Mondur CB-60	370

The outstanding attributes of this coating are its excellent resistance to abrasion, good chemical resistance, and hardness. This coating also provides a rich look of quality to wood surfaces. Coating is recommended for use where toughness, abrasion, chemical and solvent resistance are required.

Because this coating is based on an aromatic polyisocyanate, it will develop an amber or golden color upon exposure to sunlight.

Mondur CB-60 is recommended for coatings where chalking and some film yellowing are not of control importance.

No. 7

Multron R-351-65	559
Ethylglycol Acetate	110
Toluol	110
Modaflow	1
Mondur HC	220

The attributes of this coating are related to a rapid conversion of the wet film to a surface dry condition, good color stability of the film, along with resistance to chemicals and solvents. The color stability of this coating is related to the aliphatic-aromatic nature of the polyisocyanate **Mondur HC**.

This formulation should be considered in applications where small change in film color is acceptable and where the highest order of abrasion resistance is not required.

Mondur HC is recommended where good resistance to chalking and color change is required.

No. 8

Multron R-351-65	266
Multron R-211	173
Ethylglycol Acetate	208
Toluol	151
Modaflow	1
Desmodur N-75	201

This formulation is based on the light-stable polyisocyanate **Desmodur N-75**. This coating would thus be recommended for those applications where a light-stable coating is required. The coating shows good abrasion resistance, along with the chemical and solvent resistances typical of two-component urethane systems.

No. 9

Desmophen 650A	137

Multron R-221	135
Ethylglycol Acetate	283
Toluol	165
Modaflow	2
Desmodur N-75	278

This coating is light stable and has an outstanding level of resistance to chemical and solvents along with high abrasion resistance. This coating would be used in the most demanding exposures where the highest order of light-stability is required.

Fast-Sanding Primer

(Polyurethane)

Desmophen 1300	250
Zinc Stearate (20% in ethyl acetate/toluol 1:1 pbw)	71
Toluol	125
Methyl Ethyl Ketone	125
Xylol	83
Methyl Isobutyl Ketone	76
Ethylglycol Acetate	63
Baysilone OF/OR 311 (10% in toluol)	7
Desmodur IL	200

Dry-Time (25 C -60% R.H.)
(Gardner Dry Time Recorder)
    Set-to-Touch — 2.8 min
    Surface-Dry — 5.5 min
    Hard-Dry   — 10.0 min
    Mar-Free   — 22.0 min

The fast-curing characteristics of this formulation combined with low isocyanate demand makes it attractive for wood primers. The potlife is ca 16 hr with the incorporation of the zinc stearate which improves sandability. The potlife of the urethane primer without the zinc stearate is ca 30 hr. The NCO/OH ratio for this formulation is 0.65.

Emulsion Wood Primer

(Vinyl Acrylic)

	lb	gal
Water	75	9.0
Tamol 731-25	9	1.0
Tritonex 102	4	.5
Foamicide 581 B	2	.3
Ethylene Glycol	20	2.2
Talc Chemet Ruby 400	120	5.2
B.C.W.L. Oncor 45X	145	4.4
TiO_2 **Ti-Pure R-901**	115	3.4
Methocel 4000 cps Sol'n. (2%)	50	6.0

Roller mill:

	lb	gal
Methocel 4000 cps Sol'n. (2%)	200	24.0
Carbitol Solvent	17	2.0
Varaqua 928	236	26.0
*****Varkyd 515-100** Blend	67.5	8.1
Water	75	9.0

PVC (%)	38.3
Total N.V. by Wt. (%)	51.2
Vehicle N.V. by Wt. (%)	26.8
Wt. per gal (lb)	11.25
Visc. (KU)	85–95
Brushing	Excellent
Leveling	Excellent
Sagging Resistance	Excellent
Settling	None
Skinning	None
Gloss	Flat
Dry — Set-to-Touch	15–30 min
Hard	1–2 hr
Enamel Holdout	Excellent

*****VKD 515-100**

	lb	gal
Igepal CO 630	58	6.8
PMO-10	3	0.4
Cobalt Naphthenate (6%)	1.5	0.2
Lead Naphthenate (24%)	2	0.3

Fast Dry Lumber Siding Primer

(Vinyl Toluene)

	lb	gal
Basic Sulfate White Lead (**EP #41**)	200.0	3.75
Titanium Dioxide (**Titanox RA-50**)	150.0	4.29
Magnesium Silicate (**Asbestine 3X**)	140.0	6.06
Celite 281	30.0	1.53
Bentone 34	4.0	0.26
Keltrol 1074	336.0	44.19
VM&P Naphtha	188.0	30.13
Mineral Spirits	64.0	9.76
Cobalt Naphthenate (0.03%) (6%)	1.4	0.17
Exkin No. 2 (0.2%)	2.0	0.26

Visc. (KU)	70
PVC (%)	40.0
Vehicle Solids (%)	34.3
Total Solids (%)	65.1
Wt. per gal (lb)	11.08

Chapter VII

ENAMEL PAINTS

Enamels are characterized by gloss — a result of low pigment volume concentration. Depending on the degree of glossiness, the weathering requirements, the pigment load (either for aesthetic or mechanical reasons such as adding nonskid properties to a slippery concrete surface), and cost limitations, a myriad of trade-sales and industrial enamels exist for just about every need.

Semi-Gloss Enamel

(Alkyd)

	lb	gal
Titanium Calcium RCHT-X	425	15.7
RA-50	75	2.1
Thixatrol ST in Mineral Spirits (20%)	15	2.2
Mineral Spirits	33	5.0
Varkyd 701-70	270	35.0

Two passes on roller mill and add the following items:

	lb	gal
Varkyd 701-70	54	7.0
Dryfol W	16	2.0
Mineral Spirits	106	16.0
Kerosene **Amsco**	94	14.0
Cobalt Naphthenate (6%)	4	.5
Lead (24%)	8	.8
Volatile ASA	1	.1
Z-HA Safflower Oil	4	.5

PVC (%)	37.2
Vehicle Nonvolatile (Wt.) (%)	43.9
Total NV by Wt. (%)	68.8
Wt./gal (lb)	10.94
Visc. (KU)	85–90
Brushing	Excellent
Leveling	Very Good
Sagging Resistance	Excellent
Settling	None
Skinning	None
Gloss	Satin Finish
Dry — Set-to-Touch	2–3 hr
Hard	5–6 hr

If more gloss is desired, use 3 gal more mineral spirits and 3 gallons less kerosene. This may impair brushing ease and thixotropy to a slight degree.

White Semi-Gloss Enamel

(Alkyd)

	lb	gal
Ti-Pure R-900	250	7.14
Vicron	250	11.10
Thixatrol ST	6	0.70
Nuact Paste	3	0.21
Aroplaz 1254-M-70	156	19.5
Mineral Spirits	30	4.5

Grind on Cowles (Min. heat 140 F)

Aroplaz 1254-M-70	156	19.5
Aroflat 3113-P-30	200	28.00
Mineral Spirits	58	9.0
Cobalt Naphthenate (6%)	2	0.25
Lead Naphthenate (24%)	4	0.40
ASA	1.5	0.19

Wt./gal (lb)	11.16
PVC (%)	37.4
Visc. (KU)	99

Dark Green Porch and Deck Enamel

(Alkyd)

	lb/100 gal
Chrome Green A04464	149
Thixcin R	4
Haynie 121-50 Alkyd	503
Mineral Spirits	175
Cobalt (6%)	3.4
Zirconium (6%)	8
Antiskinning Agent	1
lb/gal	8.45
Visc. (KU)	77–85
Grind	6–7

High Quality Semi-Gloss Enamel

(Alkyd)

	lb/100 gal
Extended Titanium Dioxide (30% TiO_2)	335.00
Easily Dispersible Rutile Titanium Dioxide	150.00
Thixatrol ST*	6.00
Haynie 311-60 Alkyd	430.00
Mineral Spirits	176.00
Cobalt (6%)	2.58
Zirconium (6%)	4.35
Calcium (5%)	1.54
Antiskinning Agent	1.00
Post-4	6.00
lb/gal	11.06
Visc. (KU)	90–100
Grind	6
Gloss	60–70
Nonvolatile Vehicle (%)	42.5
PVC (%)	37.5
Hiding	2.5 # TiO_2

---

*Optimum performance processing range 130°–170°F.

Black Rapid-Dry Enamel

(Alkyd)

	lb/100 gal
Carbon Black	30
Thixcin R	4
Haynie 121-50 Alkyd	552
VM&P Naphtha	24
Mineral Spirits	157
Cobalt (6%)	2.8
Calcium (5%)	2.8
Lead (24%)	5.8
Antiskinning Agent	1

·lb/gal	7.5
Visc. (KU)	60–70
Grind	6–7

Gloss White Architectural Enamel

(Alkyd)

	lb	gal
Rutile Titanium Dioxide	345.1	9.87
Thixcin R	3.5	0.43
Cycopol 340-18	230.1	29.88

Disperse on roller mill and add:

	lb	gal
Cycopol 340-18	263.0	34.15
Low-Odor Mineral Spirits	160.7	24.50
Calcium Octoate Drier (5%)	5.5	0.75
Cobalt Octoate Drier (6%)	3.3	0.31
Antioxidant	0.9	0.11

Total Solids (%)	68.9
Pigment/Resin Ratio by Wt.	100/100
Visc. (KU)	75

Drier, Metal on Resin Solids:

Calcium (%)	0.08
Cobalt (%)	0.04
lb/gal	10.1

Semigloss Tint Base

(Alkyd)

	lb	gal
Rutile Titanium Dioxide	200.0	5.71
Surfex	300.0	13.30
Syntex 2964	257.0	35.00
Syntex 71	152.0	20.00
Post-4	(See Below)	
Odorless Mineral Spirits	154.0	24.44
Lead Naphthenate (24%)	2.7	0.28
Cobalt Naphthenate (6%)	1.1	0.14
Antiskinning Agent	1.0	0.13
PVC (%)	44.3	
Wt./gal (lb)	10.75	

	Control Paint (No Additive)	Paint With POST-4	Paint With Post-4
Level lb. PHG	0	5	10
Visc. (KU)	90	94	97
Sag Control (Leneta)	Passed 3	Passed 8	Passed 12

Odorless Alkyd Gloss Enamel

	lb	gal
Micro Velva A	52.00	2.35
Unitane OR-342 (TiO$_2$)	190.00	5.60
Kadox 515 (ZnO)	30.00	0.71
Thixcin R	2.95	0.23
Bekosol OP-825-70	185.00	23.85
Bekosol OP-825-70	295.00	37.95
Odorless Mineral Spirits	183.00	27.45
Nuodex Lead (24%) Odorex	4.50	0.44

	lb	gal
Nuodex Calcium (6%) **Odorex**	1.50	0.15
Nuodex Cobalt (6%) **Odorex**	5.75	0.52
Antiskinning Agent	0.75	0.75
PVC (%)	18.0	
Vol. of Solids	49.9	

Comments:

High rich permanent gloss; superb leveling; nonsagging; excellent brushability; nonyellowing; excellent coverage and mileage.

Odorless High Gloss White

(Alkyd)

	lb	gal
Titanium Dioxide	300	8.57
MPA	6	0.80
Nuact Paste	4	0.28
Calcium Naphthenate (4%)	4	0.50
Aroplaz 1257-MO-60	250	33.00
Aroplaz 1257-MO-60	250	33.00
Odorless Mineral Spirits	150	24.00
Cobalt Naphthenate (6%)	2.0	0.25
Antiskinning Agent	1.0	0.13
PVC (%)	20	
VNV (%)	46	
lb/gal	9.6	
Visc. (KU)	75–80	

Nonpenetrating Semi-Gloss

(Alkyd)

	lb/100 gal
Easily Dispersible Rutile Titanium Dioxide	136
Extended Titanium Dioxide (30% TiO_2)	300
Thixcin R	5
Haynie 121-50 Alkyd	405

	lb/100 gal
Mineral Spirits	188
Cobalt (6%)	2.6
Zirconium (6%)	6
Antiskinning Agent	1
lb/gal	10.40
Visc. (KU)	77–85
Grind	6–7

Nonpenetrating Gloss Enamel

(Alkyd)

	lb/100 gal
Easily Dispersible Rutile Titanium Dioxide	80
Extended Titanium Dioxide (30% TiO_2)	178
Thixcin R	4
Haynie 121-50 Alkyd	493
Mineral Spirits	158
Cobalt (6%)	3.2
Zirconium (6%)	8.2
Antiskinning Agent	1.4
Visc. (KU)	77–85
Grind	6–7
lb/gal	9.26

Spraying Red Automotive Enamel

(Alkyd)

	lb	gal
Bon Red Light Dupont Rt.-565D	60	3.75
Varkyd 1577-50	205	25.0
Xylol	87	12.0
MPA-60	3	0.4
Soya Lecithin	3	0.4

Grind in a pebble mill and thin with the following items:

Varkyd 1577-50	410	50.0
VM&P Naphtha	44	7.0

	lb	gal
Xylol	22	3.0
Cobalt Naphthenate (6%)	3	0.4
Manganese Naphthenate (6%)	0.5	—
Lead Naphthenate (24%)	4.5	0.5
Antiskinning Agent (Exkin #2)	1	0.1

PVC (%)	10.1
Vehicle Nonvolatile (% by Wt.)	40.2
Total N.V. by Wt. (%)	44.1
Wt./gal (lb)	8.2
Visc. (KU)	55–57
Settling	None
Skinning	None
Gloss	Very Good
Dry — Set-to-Touch (dipped)	8–10 min
Hard	3 hr

Alkyd Aerosol Enamels

FORMULA	No. 1	No. 2
	lb	gal
Ti-Pure R-900	250.0	250.0
MPA (Xylene)	4.0	4.0
Chevron 2100	490.0	415.0
Neville LX-1000	—	44.0
Xylol	220.5	251.5
Socal #3	25.0	25.0
Cobalt Naphthenate (6%)	2.5	2.5
Lead Naphthenate (24%)	4.0	4.0
Calcium Naphthenate (4%)	3.0	3.0
Exkin #2	1.0	1.0

Aerosol Packing:

Ratio — Paint to Propellant — 50:50 by weight

Propellant — Genetron 12

Actuator — 0.018 in orifice

Valve:

Stem   — 0.018 in Nylon
Body   — 0.080 × 0.020 Vapor tap
Gasket   — Neoprene
Cup   — Tinplate, unpainted with flow-in gasket
Dip tube — 6-5/16 in for 16 oz can
Spring   — Stainless steel
Supplier — Precision Valve Corp.

White Aerosol Enamel

(Alkyd)

	lb	gal
Rutile Titanium Dioxide	141.5	4.13
Aroplaz 6008-X-50	27.5	3.44
Lead Naphthenate (24%)	3.0	0.32
Ethylene Glycol Monobutyl Ether	8.5	1.15
Xylol	33.0	4.55
MPA (Xylene)	2.3	0.31

Pebble Mill Grind 24 hr

Aroplaz 6008-X-50	50.0	6.10

Add to mill base and grind an additional 30–60 min.

Aroplaz 6008-X-50	245.0	29.80
Lead Naphthenate (24%)	1.0	0.10
Cobalt Naphthenate (6%)	1.6	0.20
Manganese Naphthenate (6%)	1.6	0.20
Antiskinning Agent	0.8	0.10
Toluene	176.6	23.70
Aliphatic Solvent	176.6	29.49

lb/gal	8.40
P/B Ratio	0.86
Nonvolatile	34.80

Driers and Antioxidants (% of vehicle solids):

Pb — 0.6%   Mn—0.06%
Co — 0.06%   ASA—0.5%

	lb	gal
Aroplaz 6008 White Enamel Base	550	65.5
Propellant 12	450	40.5

lb/gal	9.44	Sprayout on tin plate[1] :
Propellant (%)	45.00	$60°$ Gloss — 90
Total Nonvolatile (%)	19.60	$20°$ Gloss — 75

[1] Using Newman Green B-14-10 Valve with B-14-1 Actuator; Vapor Pressure 39.5 psig at $70°$F

In some cases it may be necessary to add a little antisettling agent to prevent possible settling over long periods of storage. For example, it is suggested that two to three pounds of **MPA** (Xylene) be added to the grind in the White Enamel Base as insurance against settling of the titanium dioxide.

Deep Yellow Aerosol Enamel

(Alkyd)

	lb	gal
Base:		
Medium Chrome Yellow	200.0	4.06
Molybdate Orange	20.0	0.39
X-2280 IAF Compound	2.0	0.09
Post-4	8.7	1.00
Syntex 3638	350.0	42.50
Xylene	57.0	7.88

Pebble mill to 9–10 paint club

Scale reading, then add:

	lb	gal
Syntex 3638	350.0	42.50
Zirco Drier (6%)	5.9	0.79
Cobalt Naphthenate (6%)	3.0	0.36
Exkin #2	2.0	0.15
Antimar Agent	2.0	0.28
Reduction:		
Base	556.0	55.5
Toluene	322.0	44.5

	lb		gal
Reduction Visc.: 11–13 s (Ford Cup #4)			
Can Loading:			
Reduced Paint	53.4		50.7
Propellent P	46.6		49.3
PVC (%)		10.7	
Density (lb/gal)		10.0	
Vehicle Solids (%)		46.0	
Total Solids (%)		58.4	
Visc. (KU)		106	

Stipple Gloss Enamel

(Maleic Ester)

	lb/100 gal
Easy Dispersible Rutile Titanium Dioxide	105
Extended Titanium Dioxide (30% TiO_2)	316
Kadox 515 Zinc Oxide	18
Thixcin R	6
Penglo 65	416
Haynie Admiral Z-ZI	113
Mineral Spirits	106
Cobalt (6%)	7
Antiskinning Agent	1.25
Troykyd 21 BA	3.2
lb/gal	10.91
Visc. (KU)	87–97
Grind	6–7

High Solids, Light Blue Enamel

(Epoxy)

	lb	gal
Base Component:		
Vanoxy 115	522.7	55.03
Rayox R-88	122.1	3.50

	lb	gal
Sparmite	124.4	3.44
Celite 165-S	48.6	2.53
Silicone Resin SR-82	8.1	0.91
Phenol Sol'n. (50% Wt. in		
denatured ethanol)	31.4	4.14
Thixatrol ST	9.1	1.07
Phthalocyanine Blue Paste*	6.6	0.66

Curing Agent Component:

	lb	gal
Vanoxy Curing Agent H-3	232.6	28.72

Wt./gal	11.06
Usable Pot Life, hr	3½–4
Flash Point, °F Cleveland Open Cup	Above 200

---

*Phthalocyanine Blue Paste-Ball Mill Dispersion:

Vanoxy 115	15
X-2371 Monarch Blue Toner NC (Imperial)	85

Preparation of Base Component

To approximately one-half of the **Vanoxy 115** add the **Rayox R-88**, **Sparmite**, **Celite 165-S**, and the **Thixatrol ST**. Disperse by means of vigorous agitation on a Cowles Dissolver, until a temperature of 130°– 140°F is reached. Add the remaining ingredients with good agitation.

High Solids, Beige Enamel

(Epoxy)

	lb	gal
Base Component:		
Vanoxy 115	528.8	55.67
Rayox R-88	119.8	3.42
Sparmite	124.6	3.44
Celite 165-S	48.6	2.53
Red Iron Oxide N-1860	1.0	.03
Pure Yellow Iron Oxide YO-1987	1.0	.03
Silicone Resin SR-82	8.1	.91

	lb	gal
Phenol Sol'n. (50% Wt. in denatured ethanol)	31.4	4.15
Thixatrol ST	9.1	1.07

Curing Agent Component:

	lb	gal
Vanoxy Curing Agent H-3	232.8	28.75

Wt./gal (lb)	11.0
Usable Pot Life, hr	3½–4
Flash Point, °F, Cleveland Open Cup	over 200

Preparation of Base Component:

To approximately one-half of the **Vanoxy 115** add the **Rayox R-88**, **Sparmite**, **Celite 165-S**, and the **Thixatrol ST**. Disperse by means of vigorous agitation on a Cowles Dissolver, until a temperature of 130½–140 F is reached. Add the remaining ingredients with good agitation.

White Tinting Enamel .

(Epoxy)

	lb	gal

Resin Component:

	lb	gal
Epon Resin 1001-X-75	474.1	52.10
Titanium Dioxide R-900	323.9	9.50
Talc No. 399	35.4	1.60
Multifex MM	35.5	1.60
Nuosperse 657	6.3	0.80
M-P-A 60 (Xylene)	19.0	2.60
PC-1344 (60% N.V. Xylene)	2.6	0.30
Advance Antiflood Agent	4.1	0.50

Two passes over 3-roll mill

Let down:

	lb	gal
Beetle 216-8	9.4	1.10
Solvesso 150	118.5	15.90
Diacetone Alcohol	27.6	3.50

	lb	gal
Butyl Oxitol	39.5	5.30
Oxitol	39.5	5.10

Curing Agent Component:

	lb	gal
Epon Curing Agent VI-KX-60	586.6	75.20
Solvesso 150	118.3	15.90
DMP-30	7.3	0.90
Oxitol	16.3	2.10
Butyl Oxitol	33.8	4.50
Advance Antiflood Agent	4.1	0.50

Total Nonvolatiles (%)	61.8
Wt./gal (lb)	9.44
Catalyzed Visc. at 78 F (KU)	78
PVC (%)	12.6
Pigment Binder Ratio	1/1.98

Black Aerosol Enamel

(Epoxy)

	lb	gal
Base:		
Superba Beads Special	20.0	1.37
Litharge	4.0	0.05
Soya Lecithin	8.0	1.00
Post-4	8.7	1.00
Ten Cem Copper (6%)	1.1	0.15
Epi-Tex 183	200.0	25.00
Toluene	99.0	13.67

Steel Ball Mill 45–48 hr. Check grind using thin film drawdown, then add:

Epi-Tex 183	200.0	25.00

Grind 1 hr and add:

Epi-Tex 183	252.0	32.00
Lead Naphthenate (24%)	3.0	0.31

	lb	gal
Cobalt Naphthenate (6%)	2.4	0.30
Exkin #2	1.2	0.15

Reduction:

Base	356.0	44.5
Toluene	396.0	55.0
Cellosolve Acetate	4.0	0.5

Reduction Visc.	11–13 s	

Can Loading:

Reduced Paint	57.2	50.7
Propellant "P"	42.8	49.3

Visc. (KU)	88
Wt./gal (lb)	8.00
PVC (%)	3.58
Total Solids (%)	46.4
Vehicle Solids (%)	44.2

White Aerosol Enamel

(Epoxy)

	lb	gal
Base:		
Rutile TiO_2	280.0	8.00
Post-4	8.7	1.00
Soya Lecithin	8.0	1.00
Litharge	4.0	0.05
Calcium Octoate (5%)	2.5	0.33
Epi-Tex* 183	310.0	39.00

Pebble Mill to 9–10 Paint Club Scale reading, then add:

Epi-Tex* 183	309.0	38.80
Lead Naphthenate (24%)	3.0	0.31
Cobalt Naphthenate (6%)	2.1	0.26
Exkin #2	1.2	0.15
Toluene	92.0	11.10

	lb	gal
Grind 1 hr and drain mill		
Reduction		
Base	435.0	43.0
Toluene	411.0	56.5
Cellosolve Acetate	4.0	0.5

Reduction Visc. — 11–13 s on a #4 Ford Cup

Can Loading:

	% by vol	% by Wt.
Reduced Paint	62.0	59.0
Propellant "P"	38.0	41.0

Visc. (KU)	95
Wt./gal	10.2
PVC (%)	22.5
Vehicle Solids (%)	44.3
Total Solids (%)	60.4

Epoxy-Amine High Solids White Spray Coating

	lb/100 gal
Titanox RA	75.0
Cab-O-Sil	15.0
Epotuf 37-128	234.0
Cowles Disperser — approx. 5 N.S.	
Glycerin (add slowly with good agitation)	2.0
Epotuf 37-128	213.0
No. 840 Silicone	8.5
MEK	32.0
Toluol	32.0
Cab-O-Sil	5.3
Epotuf Hardener 37-611	224.0
Cowles Disperser — approx. 5 N.S.	
Glycerin (add slowly with good agitation)	1.0

	lb/100 gal
Epotuf Hardener 37-614	55.5
Ethyl Alcohol	16.0
Toluol	16.0

Uses:

Decorative and protective thick film coating for masonry, metal, and other surfaces.

Epoxy Solventless Enamel

	lb	gal
Base Component:		
Epon 815	534	56.3
TiO_2 (rutile, nonchalking type)	81	2.4
Sparmite	194	5.5
Celite 165-S	49	2.6
Thixatrol ST	10	1.2
Silicone Resin SR-82	8	0.9
Phenol Sol'n. (50% Wt. in		
Neosol Proprietary Solvent)	32	4.4
Curing Agent Component:		
Epon Curing Agent H-3	215	26.6

Wt./gal (lb)	11.2	
Usable Pot Life, hr	4–5	

Procedure:

Disperse the pigment, extenders, **Thixatrol ST** and silicone resin with about three fourths of the **Epon 815** on a Cowles Dissolver. Continue dispersion until a temperature of about 150°F is reached to insure proper development of thixotropy by the **Thixatrol ST**. The resulting dispersion should then be let down with the remaining **Epon 815** and phenol solution. The base component and curing agent component should be packaged separately.

Gloss Enamel

(Epoxy)

	lb
Rutile Titanium Dioxide[1]	100.0
Pigment Suspension Agent[2]	3.0
Wetting Agent[3]	1.0
Pigment Dispersant (25%)[4]	3.5
Morpholine	3.0
Ammonium Hydroxide (28%)	3.0
Water	75.0
Defoamer[5]	2.0
Preservative[6]	0.6

Grind — pebble mill to fineness 7 + (N.S.)

	lb
Thickener Sol'n.[7]	68.0
Defoamer[5]	4.0
Araldite PR-805, 50%	625.0
Cobalt (6%)[8]	4.0
Manganese Naphthenate (6%)[9]	2.5
Ethylene Glycol	14.0
Tint with	
Black Aqueous Pigment Dispersion[10]	18.4

Gloss, 60° Glossmeter	90+
Drying Time: Set-to-Touch	20 min
Dry hard	11–16 hr
Dry through	11–16 hr
Pencil Hardness, 24 hr	B

---

[1] Tri-Pure R-901
[2] Emultex R
[3] Triton X-100
[4] Tamol 731
[5] AF-7
[6] PMA-18
[7] Thickener Sol'n.:

	lb	gal
Water	850.0	101.5
ASE-95	122.0	13.5

Mix until well despersed and add under constant agitation:

NH$_4$OH (28%)	28.0	3.6

[8] Cyclodex Cobalt
[9] Nuodex
[10] X-2472

Semigloss Enamel

(Epoxy)

	lb
Titanium Dioxide, Rutile[1]	246.7
Dispersant[2]	2.2
Wetting Agent[3]	8.1
Antifoam Agent[4]	2.2
Water	85.8

Disperse on Cowles (or equivalent) at 3400 rpm for 20 min descrease speed and add:

Water	28.2
Propylene Glycol	43.1
Preservative[5]	1.1

Mix 5 min with good agitation and add:

Acrylic Emulsion, 46%[6]	379.8
Araldite PR-805 (50%)	174.8
Cobalt Drier (6%)[7]	0.2
Lead Drier (24%)[8]	1.2
Butyl **Cellosolve**	26.3
Thickener Sol'n. (3%)[9]	44.2

---

[1] Ti-Pure R-900
[2] Triton X-100
[3] Tamol 731
[4] Colloid 600
[5] Super Ad-It
[6] Rhoplex AC-61
[7] Advacar Cobalt
[8] Advacar Lead
[9] Hydroxyethylcellulose WP 4400

Painter's Enamel

(Epoxy)

	lb	gal
Titanox C-50	285.00	9.85
Titanium Dioxide, R-900	95.00	2.73
Zinc Oxide, Kadox 515	8.00	0.17
Nevchem 120-70	92.00	11.50
Mineral Spirits	15.00	2.24
Plaskon-3139 (70% NV)	190.00	24.00
Thixatrol ST	4.00	—
Plaskon-3893 (80% NV)	2.50	0.29
Maglite D Dispersion	24.00	2.26

Grind above on Dissolver or suitable high speed disperser to achieve a temperature of 160 F. Hold for grind of 5½ minimum, then add:

	lb	gal
Plaskon-3139 (70% NV)	198.00	25.00
Mineral Spirits	153.00	22.71
Cobalt Naphthenate (6%)	2.00	—
Calcium Naphthenate (4%)	2.00	—
Exkin #2	0.80	—
Plaskon-633	19.00	2.25

Pigment by Wt. (%)	35.6
Resin Solids by Wt. (%)	30.8
PVC (%)	24.3
Visc. overnight, at 25°C (KU)	109

Gray High Solids Epoxy-Polyamide Spraying Enamel

		lb/100 gal
A	Titanox RA	34.0
	Cab-O-Sil	22.0
	Epotuf 37-128	450.0

Disperse on high speed Cowles Disperser to 6 + N.S. fineness

Glycerin	1.8

Disperse additional 5 min

	lb/100 gal
SR-82	9.0
MEK	51.0
Toluol	51.0

		lb/100 gal
B	Superjet Lampblack	2.2
	Cab-O-Sil	5.5
	Epotuf Hardener 37-112 (EH-34)	236.0

Disperse on high speed Cowles disperser to 6 + N.S. fineness

Glycerin	0.9

Disperse additional 5 min.

Pennsalt S-2	30.0

Vanoxy Green, Enamel

(Epoxy)

	lb	gal
Base Component:		
Vanoxy 201-X-75	310.4	34.10
Rayox R-88	123.4	3.59
Chromium Oxide Green X-1134 C.P.	139.8	3.30
Ramapo Green B GP-501-D	14.0	0.99
MPA Xylene	2.4	0.33
Beetle 216-8	12.5	1.47
Modaflow	1.9	0.23
Methyl Isobutyl Ketone	39.8	5.99
Curing Agent Component:		
Vanamid 315-X-70	184.6	23.67
Xylene	55.3	7.73
Vansolve EE	122.6	15.78
Vansolve EB	21.1	2.82

Total Nonvolatile (%)	63
Pigment/Binder, Wt. Ratio	43.5/56.5
lb/gal	10.3
Visc., Stormer (KU)	69

Procedure:

Base Component — Disperse the pigments and MPA Xylene in a suitable portion of the **Vanoxy 201-X-75** and available solvent using a 3-roll mill. Let down with the remaining **Vanoxy 201-X-75**, solvent **Beetle 216-8** and **Modaflow**.

Curing Agent — Charge the **Vanamid 315-X-70** to a suitable container. Under constant agitation add the solvents and mix thoroughly.

Package the base component and the curing agent component separately. Mix just prior to use. Allow one-hour induction time before applying.

Red Epoxy Enamel

		lb
A	**Araldite 471 X-75**	279.9
	Red Iron Oxide	189.1
	MIBK	38.8
	DIBK	38.8
	Flow Control Agent	11.3
	Antisag Agent	7.6
B	**CIBA Polyamide 815 X-70**	157.9
	Solvesso 150	132.4
	MIBK	35.0
	DIBK	35.9

Epoxy Green Enamel

	lb	gal
Base Component:		
Vanoxy 201-X-75	310.4	34.10
Rayox R-88	123.4	3.59
Chromium Oxide Green X-1134 C.P.	139.8	3.30
Ramapo Green BGP-501-D	14.0	0.99
MPA 60 (Xylene)	4.0	0.55
Beetle 216-8	12.5	1.47
Modaflow	1.9	0.23
Methyl Isobutyl Ketone	39.8	5.99

	lb	gal
Curing Agent Component:		
Vanamid 315-X-70	184.6	23.67
Xylene	55.3	7.73
Vansolve EE	122.6	15.78
Vansolve EB	21.1	2.82

Procedure:

Base component — Disperse the pigments and **MPA 60** (Xylene) in a suitable portion of the **Vanoxy 201-X-75** and available solvent using a 3 roll mill. Let down with the remaining **Vanoxy 201-X-75**, solvent **Beetle 216-8** and **Modaflow**.

Curing Agent — Charge the **Vanamid 315-X-70** to suitable container. Under constant agitation add the solvents and mix thoroughly.

Package the base component and the curing agent component separately. Mix just prior to use. Allow 1 hr induction time before applying.

Epoxy White Enamel

	lb	gal
Pebble Mill		
A **Epon Curing Agent V-15-X-70**	169.1	21.13
Titanium Dioxide R-610	295.8	8.44
Talc No. 399	29.6	1.31
MPA 60 (Xylene)	9.9	1.35
Cyclo-Sol 53	72.4	9.86
Oxitol	44.4	5.72
Mill Rinse:		
IPA (99%)	57.8	8.84
Cyclo-Sol 53	72.4	9.86
DMP-30	1.2	0.15
B **Epon Resin 1001-CX-75**	294.8	32.15
Beetle 216-8	9.8	1.18

Total Nonvolatile (% Wt.)	65.8
Wt./gal (lb)	10.6
Pigment/Binder (by Wt.)	0.88/1.0
Visc. of A & B (KU)	62

Yellow Gloss Epoxy Enamel

(Polyester)

	lb	gal
Pigment Dispersion:		
Aroflint 606 Component	212	25.23
MPA 60 (Mineral Spirits)	15	2.21
Flow Control Agent	5	0.68
Medium Chrome Yellow	300	6.42
Solvent	109	15.50
Clear Activator:		
Aroflint 252 Component	433	46.34
Solvent	26	3.70
Enamel Formula:		
Pigment Dispersion	641	50.04
Clear Activator	459	50.04

Gloss White Enamel

(Polyester)

	lb	gal
Pigment Dispersion:		
Titanium Dioxide	262.0	7.91
MPA (Xylene)	10.0	1.47
Flow Control Agent	4.5	0.61
Aroflint 606 Component	100.0	11.90
Solvent	11.0	1.55
Roller mill or pebble mill:		
Aroflint 606 Component	112.0	13.33
Solvent	94.0	13.27
Clear Activator:		
Aroflint 252 Component	433.0	46.34
Solvent	26.0	3.70
Enamel Formula:		

	lb	gal
Pigment Dispersion	593.5	50.04
Clear Activator	459.0	50.04

PVC (%)	14.5	
Total Solids (%)	70.2	
Visc. (KU)	60–65	
Brushing	Good	
Sag and Flow Control	Good	
Ratio of Components of **Aroflint 252/**		
Aroflint 606 solids	55.45	

Blue Gloss Enamel

(Polyester)

	lb	gal
Pigment Dispersion:		
Aroflint 606 Component	212.0	25.23
MPA 60 (Mineral Spirits)	10.0	1.47
Flow Control Agent	4.5	0.61
Titanium Dioxide	162.0	4.99
Phthalocyanine Blue	18.0	1.57
Solvent	114.3	16.17
Clear Activator:		
Aroflint 252 Component	433.0	46.34
Solvent	26.0	3.70
Enamel Formula:		
Pigment Dispersion	520.8	50.04
Clear Activator	459.0	50.04

PVC (%)	12.5	
Total Solids (% by Wt.)	66.9	
Total Solids (% by vol.)	52.8	
Visc. (KU)	60–65	
Ratio of Components of **Aroflint 252/**		
Aroflint 606 solids	55.45	

Yellow Industrial Enamel

(Alkyd)

	lb	gal
Yellow Iron Oxide	120	3.56
Titanium Dioxide	80	2.29
Bentone 38	4	0.30
4% Calcium Naphthenate	5	0.60
Varkyd 1577-50HS	550	69.70
SoCal 2	60	8.25
VM&P Naphtha	60	9.50
Cobalt Naphthenate (6%)	2.0	0.25
Zirco Naphthenate (6%)	1.5	0.21
Exkin #2	1.0	—
PVC (%)		18.0
Vehicle N.V. (% Wt.)		56.0
Total N.V. (Wt.)		66.0
Wt. per gal		9.25 lb
Visc. (KU)		67–77

Automotive and Equipment White Enamel

(Alkyd)

	lb	gal
Titanox RANC	200	5.72
Litharge	2	—
Varkyd 1577-50 HS	500	63.50
Bentone 38	5	—
Aromatic Solvent	208	29.35
Cobalt Naphthenate (6%)	2	—
Calcium Naphthenate (4%)	1.5	—

Bright Red Automotive Enamel

(Alkyd)

	lb	gal
Cyanamid Bonadur Red 20-6485	60	0.41
Cyanamid Valencia Orange 40-8340	90	1.89

	lb	gal
Bentone 38	5	0.33
Calcium Naphthenate (4%)	5	0.60
Varkyd 1577-50HS	580	73.40
SoCal 2	64	9.00
VM&P Naphtha	64	10.00
Cobalt Naphthenate (6%)	2.0	0.25
Zirco Naphthenate (6%)	1.5	0.21
Exkin #2	1.0	—
PVC	9%	
Vehicle N.V. (Wt.)	41%	
Total N.V. (Wt.)	51.5%	
Wt. per gal	9.1 lb	
Visc. (KU)	67–77	

Industrial Orange Enamel

(Alkyd)

	lb	gal
DuPont YE-689-D Molybdate		
Orange	136	2.86
Rutile Titanium Dioxide	10	0.29
Bentone 38	5	0.33
Calcium Naphthenate (4%)	5	0.60
Varkyd 1577-50HS	628	79.50
SoCal 2	99	13.75
Cobalt Naphthenate (6%)	2.25	0.27
Zirco Naphthenate (6%)	1.75	0.24
Exkin #2	1.00	—
PVC (%)	10.5	
Vehicle N.V. (% Wt.)	43.5	
Total N.V. (% Wt.)	53.0	
Wt. per gal	9.1 lb	
Visc. (KU)	67–77	

Chapter VIII

INTERIOR PAINTS

Flat Wall Paints

The important characteristics of a good flat wall paint are: high hiding power; a uniform appearance, resistance to burnishing, staining, scrubbing, chipping, and yellowing; good pigment binding properties; hard and smooth film; and moderate cost.

Alkyd Flat Wall Paints

These paints are relatively expensive to produce but are preferred by professional painters for one-coat effectiveness on hard-wear surfaces. The ideal alkyd wall paint should have no low-angle sheen so that surface irregularities are obscured.

Formula No. 1

Acintol FA1 or **FA2** Tall Oil Fatty Acid	181 g
Phthalic Anhydride	190 g
Trimethylolethane (TME)	155 g

Procedure: Fusion Cook

Charge all ingredients to a reactor fitted with thermometer, stirrer, inert gas, and open inlets. Adjust inert gas flow to 0.2 cfm/ga. Heat to 245 C. and hold for acid number of 7–9. Cool and reduce with mineral spirits. Approximate cooking time: 5.5–6 hr.

	Acintol FA1	Acintol FA2
Acid Number (Solids)	7–9	7–9
Color, Gardner, 1963 (30% nonvolatile)	7	4
Visc. Gardner/Holdt, at 25°C		
30% (nonvolatile)	Z_4	Z_1
25% (nonvolatile)	Y	P

No. 2

Acintol D50LR Distilled Tall Oil	640 g
Phthalic Anhydride	296 g
Pentaerythritol	85 g
Glycerol (95%)	150 g

Procedure: Fusion Cook

Charge all ingredients to a 2-l reactor fitted with stirrer, thermometer, inert gas, and 18 in air condenser. Heat to 210 C and hold for 2 hr. Run to 240 C and hold for acid number 12–15. Cool and reduce with odorless mineral spirits to 30% nonvolatile.

Nonvolatile (%)	30
Acid Number (solids)	12–15
Visc. Gardner/Holdt, (30% nonvolatile) at 25 C	N–O
Color, Gardner, 1963	5

No. 3

Acintol D29LR Distilled Tall Oil	660 g
Phthalic Anhydride	444 g
Pentaerythritol (technical)	246 g
Ethylene Glycol	93 g

Procedure: Fusion Cook

Charge all ingredients to a four neck 2-ℓ reactor fitted with stirrer, thermometer, and inert gas and 18 in. air condenser. Adjust gas flow at 0.1 cfm/gal. Heat to 210 C. and hold for 1 hr. Then heat to 240 C. and hold for acid number of 10–15. Cool and reduce with odorless mineral spirits to 30% nonvolatile.

| Nonvolatile (%) | 30 |

Acid Number — (Solids)	10–15
Visc., Gardner/Holdt, (30% nonvolatile) at 25 C	M–P
Color, Gardner, 1963	3–4

Solvent cooks of the above formulation can be made with adjustment of the ethylene glycol content. It should be noted that ethylene glycol will azeotrope off with the water and Xylene.

	No. 4	No. 5	No. 6	No. 7	No. 8	No. 9	No. 10	No. 11
Acintol FA2 Tall Oil Fatty Acid	—	553	—	502	—	452	—	384
Acintol FA1 Tall Oil Fatty Acid	555	—	504	—	454	—	385	—
TME (Tri-methylol-ethane)	310	312	355	356	386	388	442	444
Phthalic Anhydride	400	400	437	437	475	475	525	525
Xylene	64	64	65	65	66	66	66	67

Procedure:

Charge all ingredients into a 2-ℓ reactor equipped with stirrer, thermometer, inert gas inlet, reflux condenser, and azeotrope receiver. Adjust inert gas flow to 0.2 cfm/gal. Heat to 245 C. in approximately 2 hr and hold for an acid number of 10 or less. Remove condenser assembly, adjust inert gas to 0.4 cfm/gal. Cool to 140 C and reduct to 60% solids with Xylene.

	No. 12 (Low Odor)		No. 13 (Odorless)		No. 14 (Odorless)	
	lb	gal	lb	gal	lb	gal
TiPure R-900	200	5.8	200	5.8	—	—
Titanox C-50	—	—	—	—	400	13.7
Atomite	300	13.4	375	16.7	—	—

| | No. 12 | | No. 13 | | No. 14 | |
| | *(Low Odor)* | | *(Odorless)* | | *(Odorless)* | |
	lb	gal	lb	gal	lb	gal
Duramite	200	8.9	200	8.9	400	17.8
Thixcin-R	—	—	3	0.4	3	0.4
Aroflat 3113-P-30	441	61.0	—	—	—	—
Aroflat 3050 MO	—	—	334	48.0	334	48.0
Cobalt Odorex (6%)	1	0.1	1	0.1	1	0.1
Lead Odorex (24%)	2	0.2	2	0.2	2	0.2
Antiskinning Agent	1	0.1	1	0.1	1	0.1
Low Odor Mineral Spirits	66	10.5	—	—	—	—
Odorless Mineral Spirits	—	—	125	19.8	124	19.7
PVC (%)	66.7		67.0		67.3	
Visc. (KU)	Thixo. body		85–90		85–90	
Paint N.V. by vol. (%)	42.1		47.0		47.2	
Paint N.V. by Wt. (%)	69.0		72.8		73.8	
Vehicle N.V. by Wt. (%)	26.8		27.6		27.6	

| | No. 15 | | No. 16 | | No. 17 | |
| | *(Regular)* | | *(Low Odor)* | | *(Low Odor)* | |
	lb	gal	lb	gal	lb	gal
TiPure R-900	200	5.8	200	5.8	—	—
Titanox C-50	—	—	—	—	400	13.7
Atomite	300	13.4	300	13.4	—	—

	No. 15 (Regular)		No. 16 (Low Odor)		No. 17 (Low Odor)	
	lb	gal	lb	gal	lb	gal
Duramite	200	8.9	200	8.9	325	14.5
Thixcin-R	3	0.4	3	0.4	3	0.4
CAF	442	61.0	–	–	–	–
CAF-10	–	–	442	61.0	442	61.0
Glyceryl-Mono-Oleate	–	–	–	–	4	0.5
Cobalt Naphthenate (6%)	1	0.1	–	–	–	–
Cobalt Odorex (6%)	–	–	1	0.1	1	0.1
Lead Naphthenate (24%)	3	0.3	–	–	–	–
Lead Odorex (24%)	–	–	3	0.3	3	0.3
Antiskinning Agent	1	0.1	1	0.1	1	0.1
Mineral Spirits	67	10.0	–	–	–	–
Low Odor Mineral Spirits	–	–	63	10.0	59	9.4
PVC (%)	67.1		67.1		67.1	
Visc. (KU)	85–95		85–95		85–95	
Paint N.V. by vol. (%)	42.2		42.2		42.3	
Paint N.V. by Wt. (%)	68.7		68.7		69.3	
Vehicle N.V. by Wt. (%)	26.1		26.1		26.2	

Latex Flat Wall Paint

Practically all interior latex paints are based on three emulsified polymers: styrene-butadiene, polyvinyl acetate, and acrylic polymers.

Styrene-butadiene is resistant to alkalies and water. The advantage of this type is its low cost; the disadvantage is that as the film ages, it continues to cure becoming brittle and yellow.

Polyvinyl acetate is easier to manufacture but it is more expensive to produce and it is highly sensitive to water and alkalies.

Acrylic polymers reflect ultraviolet light and are resistant to discoloration. They are also more resistant to water and alkalies than the polyvinyl acetates even though they are based on esters.

FORMULA NO. 1

(One Coat)

	lb	gal
Pigment Grind:		
Water	290.0	35.0
Igepal CTA-639	4.0	0.4
Tamol 731	7.0	0.8
Ethylene Glycol	30.0	3.2
Nilfoam 7	2.0	0.2
Cellosize Hydroethyl Cellulose		
QP-4400H	5.0	0.5
TiPure R-901	300.0	9.0
ASP 170	50.0	2.3
Satintone No. 1	50.0	2.3
Barytes No. 1	50.0	1.3
Dowicil 75	1.0	—
Let Down:		
Ucar Latex 365	360.0	39.6
Ucar Filmer 351	6.0	0.8
Dibutyl Phthalate	6.5	0.8
Nilfoam 7	2.5	0.3
Water	29.0	3.5
PVC (%)	42.5	
Total Solids (%)	55.6	
Angular Sheen (85°)	14–18	
Visc. (KU)	85–90	
Wt. per gal (lb)	11.9	
Brightness	92	
Contrast Ratio (untinted)	0.98	

No. 2

	lb	gal
Water	200	24.0
Tamol 731-25	6	.7
Foamicide KP-140	3	.4
Ethylene Glycol	20	2.1
Odorless Mineral Spirits	3.1	.5
Texanol	8	1.0
Whiting Snowflake	100	4.45
Microcel T-38	25	1.3
Titanium Dioxide Ti-Pure R-901	275	8.25
Cellosize QP-15000 Sol'n. (2.5%)	75	9.0
Talc Fibrene C-500	75	3.3
Mildewcide Metasol 57	0.5	—
Cellosize QP-15000 Sol'n. (2.5%)	100	12.0
Varaqua 928	264	29.0
Water	50	6.0

PVC (T)	53.0
Total N.V. by Wt. (%)	52.3
Wt. per gal (lb)	11.5
Visc. (KU)	85–90
Application Properties	Excellent
Leveling	Very Good (outstanding for this class of latex paints)
Gloss	Flat
Color Uniformity	Excellent
Recoat Uniformity	Excellent
Sheen Uniformity	Very Good

No. 3

(White)

Water	500
KTPP	1
Tamol 731	5
Tergitol NPX	1
PMA 30	0.50
Ti-Pure R-901	150

Camel Carb	200
Burgess Opti White	125
Nytal 300	50
Ultra Blue	0.75
Ethylene Glycol	10
Carbitol Acetate	10
Natrosol 250 MR	5
Defoamer	2
Parco 37-C-55	150
Yield (gal)	100
PVC (%)	71.3
Solids (%)	53.4
Visc. (KU)	85
lb/gal	12.1

No. 4

(White)

	lb	gal
Water	150.0	18.0
Super Ad It	0.3	—
Nopcosant K	8.0	0.8
TKPP	0.3	—
Hyponic PE-90	4.0	0.4
Ethylene Glycol	20.0	2.2
Titanox RA-50	120.0	3.5
Optiwhite Clay	165.7	9.0
Barytes FFF703	88.0	2.4
Atomite	164.0	7.3
Foamaster G	2.0	0.2

Grind with a Cowles-type mixer for 15–20 min (3800–4500 fpm).

Add the following at lower speed:

Foamaster G	2.0	0.2
Texanol	10.0	1.2
Cellosize WP-4400 (2½% sol'n.)	226.7	27.2
Geon 450 x 20	141.3	15.0
Water	125.9	15.1

PVC (%)	75.0
Total Solids (%)	51.2
Wt. gal (lb)	11.9
Nonvolatile, vol. (%)	29.3
Visc. (KU)	85–90

No. 5

	lb	gal
Water	200.0	24.00
Daxad 30 (25%)	7.0	0.72
Dowicil 100	1.0	0.10
Rutile Titanium Dioxide (Ti-Pure R-911)	170.0	5.06
Antifoamer (50%) (XR-62149)	2.0	0.25
Clay (ASP 170)	50.0	2.31
Calcined Clay (Satintone #1)	50.0	2.28
Calcium Carbonate (Snowflake)	65.0	2.89
Silica, Amorphous (Gold Bond R)	50.0	2.49
Methocel NC-1214-5 (10%)	10.0	1.17
Ethylene Glycol	15.0	1.62

Grind above materials on high speed dispersion equipment. Add following materials and mix until smooth.

Water	204.9	24.64
Micro Cel T-70	30.0	1.50
Methocel NC 1214.5 (10%)	45.0	5.27

Let Down:

Dow Latex 307 (48%)	213.5	25.30
Defoamer (913 BL)	4.0	0.40

PVC (%)	58.0
Solids: Wt. (%)	47.2
vol. (Pigments and Binder)	28.5
Visc. (KU)	85–90
pH, adjust to	9.0–9.5

Polyvinyl Acetate Interior Flat White

(lb/100 gal)

FORMULA	No. 1	No. 2	No. 3	No. 4	No. 5
Water	←		100.0		→
Cellulosic Thickener (3%)	←		100.0		→
Lecithin	←		3.0		→
Wetting Agent	←		3.0		→
Dispersant	←		5.0		→
Ethylene Glycol	←		20.0		→
Antifoam	←		1.0		→
Titanium Dioxide	←		150.0		→
Silicate Pigment	←		30.0		→
Kaolin Pigment	←		100.0		→
Talc Pigment	125.0	72.5	20.0	72.5	20.0
Min-U-Sil 10		50.0	100.0		
Min-U-Sil 30				50.0	100.0
PMA 18	0.3				

Dispersed with high speed impeller mixer

	No. 1	No. 2	No. 3	No. 4	No. 5
Water	172.5	165.0	155.0	167.5	155.0
Cellulosic Thickener	127.5	135.0	145.0	132.5	145.0
Carbitol Acetate	←		9.0		→
Antifoam	←		2.0		→
PVAc-Maleate Latex	←		175.0		→
PVC (%)	←		61.5		→
Visc. (KU)					
Initial	90	88	87	88	88
Freeze Thaw	103	101	102	102	100
2 weeks at 120 F	105	101	98	102	100
2 months at 77 F	102	98	95	98	96
Fineness of Grind —					
Hegman	3½	4½	5	4½	5
Package Stability*					
2 weeks at 120 F	←		Excellent		→
2 months at 77 F	←	Excellent to very good			→

*Includes liquid separation, settling and ease of remixing.

	No. 6	No. 7	No. 8
		lb/100 gal	
Water	100.0	100.0	100.0
Cellulosic Thickener (2%)	100.0	100.0	100.0
Ethylene Glycol	25.0	25.0	25.0
Wetting Agent	1.5	1.5	1.5
Preservative	0.5	0.5	0.5
Antifoam	1.0	1.0	1.0
Min-U-Gel 500	0	5.0	15.0
KTPP	1.0	1.0	1.0
Organic Dispersant (25%)	4.0	4.0	4.0
Titanium Dioxide	210.0	210.0	210.0
Water	–	43.5	75.0
Kaolin (+90 brightness)*	75.0	69.7	58.9
Calcium Carbonate Pigment	120.0	120.0	120.0
Silica Pigment	30.0	30.0	30.0

Above dispersed with high speed impeller mixer.

	No. 6	No. 7	No. 8
Water	122.0	98.5	97.0
Cellulosic Thickener (2%)**	75.0	55.0	25.0
PVAc Latex	280.0	280.0	280.0
Wetting Agent	2.5	2.5	2.5
Antifoam	1.5	1.5	1.5
Dibutyl Phthalate	5.0	5.0	5.0
Coalescent	5.0	5.0	5.0

*Min-U-Gel 500 was substitued for an equal volume of kaolin pigment.
**The final addition of cellulosic thickener was made to adjust paint viscosity to 85–90 KU.

Butadiene-Styrene Flat Latex Paint

FORMULA NO. 1

(White)

	lb	gal
Add in the listed order:		
Water	365.0	43.80
Preservative	1.0	.10
Defoamer	2.0	.31
Freeze-Thaw Stabilizer	15.0	1.61
Hydroxy Ethyl Cellulose	5.0	.42
Dispersing Agent	4.6	.50
Titanox RCHT X	550.0	20.30
Micro Cel T-38	25.0	1.05
Celite 281	25.0	1.31

Grind the above in a high speed mill and let down at a slower speed with:

	lb	gal
Wetting Agent	4.3	.50
Styrene Butadiene	253.5	30.00
Cobalt Naphthenate (6%)	.8	.10
Pigment (%)	47.9	
Vehicle (%)	52.1	
Vehicle N.V. (%)	19.4	
PVC (%)	61.5	

NO. 2

(42% PVC)

	lb
Grind:	
Water	121.0
Potassium Tripolyphosphate	1.5
Tergitol NPX	2.0
Ethylene Glycol	15.0
Rutile **Zopaque R-55** Titanium Dioxide	200.0

	lb
Calcium Carbonate (**Camel Carb**)	89.0
Calcined Clay (**Satintone #1**)	100.0
Celite 281	25.0
Hydroxyethylcellulose (**QP-4400**, 3% Sol'n.)	25.0

Disperse using high speed disperser adding in above order.

Let Down:

	lb
Water	46.0
Colloids 581B	.4
PMA 30	.2
Hydroxyethylcellulose (**QP-4400**, 3% Sol'n.)	107.0
Latex (**Dow 300**) 47% Nonvolatile	393.0

Procedure:
    Add in order with agitation.

No. 3

	lb/100 gal	kg/100 lb
Grind:		
Water	200	24
Pigment Dispersant (25%)	8	.960
Rutile Titanium Dioxide	200	24
Dow Polyglycol P 1200	3	.360

Slurry with agitation for 5 min

	lb/100 gal	kg/100 lb
Calcinated Clay	10	1.2
Calcium Carbonate	114	13.6

Mix for 15–30 min

Let Down:

	lb/100 gal	kg/100 lb
Defoamer (50%)	3	.360
Methocel K15M† } added as a sol'n.	3.7	.444
Water } (2%)	181.3	21.8
Dow Latex 308 (54%)	290	34.7

	lb/100 gal	kg/100 lb
Ethylene Glycol	15	1.8
Dowicil 75 Preservative	1.25	.150

Mix:

PVC (%)	50
Nonvolatile Contents (%)	50
pH	Adjust to 9.0
Visc. (KU)	85–90

---

†**Methocel J20MS, J12MS, K15MS,** may be substituted. If so, the pH must be above 8.5 to completely break the cross link needed for good dispersion.

No. 4

Addition of 10% **Carboset 514** (replacing a like amount of latex solids) to butadiene-styrene flat paint improves brushability with fewer lap marks, color development, resistance to water spotting, adhesion and stain removal.

	lb
Methocel 90HG, 15,000 cps, (2% Sol'n.)	100.0
Water	71.0
Igepal CO-630	3.0
Ammonium Hydroxide	2.0
Polyglycol P-1200	3.0
Carboset 514	29.0
TiPure R-901	200.0
Satintone #1	110.0
Snowflake Whiting	180.0

Disperse above using high speed impeller; let down with the following, using moderate agitation:

Water	161.1
Methocel 90HG, 15,000 cps, (2% Sol'n.)	160.0
Butadiene-Styrene Latex (47% solids)	166.5
Surfynol 1045	3.0
Ethylene Glycol ⎱ Premix	15.0
PMA-30 ⎰	1.0

	lb
Ammonium Hydroxide (28%)	.7
PVC (%)	65.0
Nonvolatile (%)	50.0
Wt./gal (lb)	11.6

	No. 5 (White)		No. 6 (White)	
	lb	gal	lb	gal
Ti-Pure R-901	200	6.0	200	6.0
Driwalite	450	20.0	450	20.0
Daxad 30	5	0.5	5	0.5
Colloid 606	2	0.3	2	0.3
Ethylene Glycol	28	3.0	28	3.0
Troysan PMA-30	.3	—	.3	—
Igepal CO-610	2	0.2	2	0.2
Natrosol 250 H.R. (3%)	92	11.0	92	11.0
Water	183	22.0	175	21.0
Genflo 355 (48 N.V.)	312	37.0	—	—
Dow 300 (47 N.V.)	—	—	323	38.0
PVC (%)	59.6		59.7	
Visc. (KU)	80–85		80–85	
Paint N.V. by vol. (%)	46.6		46.1	
Paint N.V. by Wt. (%)	63.1		63.2	
Vehicle N.V. by Wt. (%)	24.4		24.5	

Acrylic Paints

FORMULA	No. 1	No. 2	No. 3
	lb/100 gal		
Water	150	150	150
Cellulosic Thickener (2½%)	125	125	125
Organic Disperstant (25%)	10	10	10
Antifoam	1	1	1
Coalescent*	15	15	15
Ethylene Glycol*	30	30	30
Preservative*	0.5	0.5	0.5

	lb/100 gal		
Titanium Dioxide	210	210	210
Min-U-Sil 30	85	85	85
Kaolin Pigment**	133.3	127.9	116.7
Min-U-Gel 500	0	5	15

Above dispersed with a high speed impeller mixer.

Water	72.5	85.0	133.0
Cellulosic Thickener (2.5%)***	160.5	148.0	100.0
Antifoam	2.0	2.0	2.0
Acrylic Latex	245.0	245.0	245.0
Total Yield (gal)	111.6	111.6	111.6
PVC (%)	54	54	54

*Added as a premix.
**Min-U-Gel 500 was substituted for an equal volume of the kaolin pigment.
***The final addition of cellulosic thickener was made to a viscosity of 93–100 KU.

	No. 4 (White)	No. 5 (White)	No. 6 (White)	No. 7 (White)	No. 8 (White)
Water	←————————111.2————————→				
Cellulosic Thickener (2.5%)	←————————100.0————————→				
Dispersant	←————————10.0————————→				
Wetting Agent	←————————2.0————————→				
Hexylene Glycol	←————————30.0————————→				
Ethylene Glycol	←————————15.0————————→				
Antifoam	←————————1.0————————→				
Titanium Dioxide	←————————180.0————————→				
Silicate Pigment	←————————27.0————————→				
Kaolin Pigment	←————————80.0————————→				
Calcium Carbonate	100.0	50.0		50.0	
Min-U-Sil 10		50.0	100.0		
Min-U-Sil 30				50.0	100.0
Preservative	1.0				

Dispersed with high speed impeller mixer

	No. 4 (White)	No. 5 (White)	No. 6 (White)	No. 7 (White)	No. 8 (White)
Water	←—————————— 50.0 ——————————→				
Cellulosic Thickener (2.5%)	←—————————— 141.2 ——————————→				
Antifoam	←—————————— 2.0 ——————————→				
Acrylic Latex	←—————————— 254.7 ——————————→				
PVC (%)	←—————————— 57 ——————————→				
Visc. (KU)					
Initial	98	98	97	98	98
Freeze Thaw	103	103	102	103	103
2 weeks at 120°F	98	97	97	97	99
2 months at 77°F	99	99	97	99	97
Fineness of Grind —					
Hegman	6	6	6	5½	5½
Package Stability*					
2 weeks at 120 F	←—————————— Excellent ——————————→				
2 months at 77 F	←—————————— Excellent ——————————→				

---

*Includes liquid separation, settling and ease of remixing.

No. 9

(White)

Water	250
Latekoll D (8%)	75
Shanco 4100-40	50
Ethylene Glycol	26
Tipure R 901	125
Atomite	400
Microcel T-70	20
PMA-30	0.2
Deefo 97-2	1.0

Add in order listed and mix on a high speed disperser for 10 min, then let down as follows:

Texanol	6.0
Deefo 97-2	1.0
Concentrated Ammonia	2.0
Acronal 290 D	120
Water	161
PVC (%)	70.4
Nonvolatile (%)	51.0
Nonvolatile by vol. (%)	30.6
Wt./gal (lb)	11.9
Visc. (KU)	85±5
pH	8.8–9.3

No. 10

(White)

Water	233.0	28.00
Ethylene Glycol	30.0	3.20
Texanol	15.0	1.90
Pigment Dispenser A	12.0	1.40
Latekoll D	17.0	2.01
Cosan PMA-30	0.3	—
Titanox RA-17	200.0	5.70
Microcel T-70	20.0	1.10
Drikalite	100.0	4.40
Gold Bond R	150.0	6.80
ASP-400	100.0	4.60

Grind on high speed disperser:

Foamaster R	2.0	0.30
Acronal 290 D	270.0	31.00
Water	83.0	9.96
Ammonium Hydroxide	2.0	0.30
PVC (%)	58.1	
Nonvolatile (%)	57.0	
Nonvolatile by vol. (%)	38.0	
Wt./gal (lb)	12.26	

	No. 11 (White Emulsion)		No. 12 (White Emulsion)	
	lb	gal	lb	gal
TiPure R-901	200	6.0	200	6.0
Drikalite	450	20.0	450	20.0
Daxad 30	5	0.5	5	0.5
Colloid 606	2	0.3	2	0.3
Ethylene Glycol	28	3.0	28	3.0
Hexylene Glycol	7	1.0	7	1.0
Troysan PMA-30	.3	–	.3	–
Igepal CO-610	2	0.2	2	0.2
Natrosol 250 H.R. (3%)	75	9.0	75	9.0
Water	167	20.0	167	20.0
Rhoplex AC-34 (46 N.V.)	356	40.0	–	–
Polyco 2719 (46 N.V.)	–	–	344	40.0
PVC (%)	60.3		60.3	
Visc. (KU)	80–85		80–85	
Paint N.V. by vol. (%)	45.1		44.5	
Paint N.V. by Wt. (%)	63.2		63.2	
Vehicle N.V. by Wt. (%)	26.7		25.2	

No. 13

Modified all-acrylics have improved leveling, color acceptance, sheen uniformity and scrub resistance.

	lb
Water	101.7
Cellosize QP-4400 (2% Sol'n.)	100.0
Triton CF-10	2.0
Carboset 514	31.0
Ammonium Hydroxide	2.0
Propylene Glycol } Premix	20.0
Colloid 600 }	2.0
TiPure R-901	152.0
Glomax LL	175.0
Snowflake Whiting	100.0
Super Ad-It	0.5

Disperse above using high speed impeller; let down with following, using moderate agitation:

Water	40.0
Cellosize QP-4400 (2% Sol'n.)	223.7
Colloid 600	2.0
Acrylic Resin Emulsion (46% Solids)	181.0
Ammonium Hydroxide (28%)	5.0
PVC (%)	64.3
Nonvolatile Content (%)	46.5
Wt./gal (lb)	11.2

No. 14

(Dripless)

Water	250.0	30.01
Latekoll D (8%)	80.0	9.47
Daxad 30	6.0	.63
Igepal CO-630	4.0	.45
TKPP	0.5	.06
PMA-30	0.3	—
Texanol	15.0	1.89
Mineral Spirits	10.0	1.57
Titanox RA-47	225.0	7.11
Micro-cel T-70	20.0	1.06
Duramite	100.0	4.43
ASP-400	100.0	4.66

Add in order listed and mix on a high speed disperser for 10 min, then let down as follows:

Concentrated Ammonia	4.0	.60
Deefo 97-2	2.0	.14
Ethylene Glycol	30.0	3.24
Acronal 290 D	251.4	28.90
Water	48.2	5.79
PVC (%)	53.0	
Nonvolatile (%)	50.9	
Nonvolatile by vol. (%)	32.6	

Wt./gal (lb)	11.5
Visc. (KU)	86±3
pH	9.3

Vinyl-Acrylic Emulsion Paints

	No. 1		No. 2	
	(White Emulsion)		*(White Emulsion)*	
	lb	gal	lb	gal
TiPure R-901	200	6.0	200	6.0
Drikalite	450	20.0	450	20.0
Daxad 30	5	0.5	5	0.5
Colloid 606	2	0.3	2	0.3
Ethylene Glycol	28	3.0	28	3.0
Hexylene Glycol	7	1.0	7	1.0
Troysan PMA-30	0.3	–	0.3	–
Igepal CO-610	2	0.2	2	0.2
Natrosol 250 H.R. (3%)	67	8.0	67	8.0
Water	224	27.0	224	27.0
Natural Starch **Resyn 2243** (55 N.V.)	310	34.0	–	–
Ucar 180 (55 N.V.)	–	–	306	34.0
PVC (%)	60		60	
Visc. (KU)	80–85		80–85	
Paint N.V. by vol. (%)	46.7		45.8	
Paint N.V. by Wt. (%)	63.8		63.5	

No. 3

(White)

	lb	gal
Premix:		
Water	250.0	30.00
AMP-95	4.5	0.57
Potassium Tripolyphosphate	1.0	–
Disperse pigments:		

	lb	gal
Unitane OR-572 Titanium Dioxide	250.0	7.50
Duramite Calcium Carbonate	100.0	4.43
Glomax LL Calcined Clay	75.0	3.50
Lorite Calcium Carbonate/Silica	75.0	3.50

Grind; then add:

	lb	gal
Ethylene Glycol ⎫ Premix	20.0	2.16
Preservative ⎭	0.5	0.05
Dibutyl Phthalate	16.0	1.85
Cellosize QP-4400 (Thickener		
2½% sol'n.)	150.0	18.00
Polyco 2151 Vinyl Acetate/		
Acrylic Copolymer Emulsion		
(55%)	228.0	25.05
Defoamer	2.0	0.26
Triton X-100 Surfactant	1.0	0.12
Water	25.1	3.01

Visc. (KU) at 77° – Fresh	73
Overnight	73
1 month	75
1 month at 125 F	77
pH – Fresh	9.5
1 month at 125°F	8.5
PVC (%)	58.0
Total Solids (% by Wt.)	52.0

No. 4

(White)

Water	125
Tamol 731	10
Triton CF-10	2
Dapro 881-S	1
Hexylene Glycol	30
Cosan 171-S	1
Zopaque R-77	200
Optiwhite	125
Celite 281	25
Asbestine 3X	75

Mix on a high speed mill and add:

Ethylene Glycol	20
Natrosol 250 MR	6
ParCryl 400	222
Dapro 881-S	1
Water	230
Yield (gal)	100
PVC (%)	59
Solids (%)	50
Visc. (KU)	84–90

No. 5

	lb	gal
Water	208	25.0
Tamol 731-25	9	1.0
Triton X-102	4	.5
Ethylene Glycol	14	1.5
Natrosol 250 HR	6	.7
Titanium Dioxide **Ti-Pure R-900**	150	4.5
Whiting **Vicron 45-3**	200	8.9
Talc **Fibrene C-500**	100	4.3
Water	142	17.0
Micro-Mica C-1000	50	2.2
Foamicide 581B	3	.4
Water	200	24.0
Mildewcide **PMA-18**	0.3	–
Soya Lecithin **Kelecin 1081**	2	.3
Texanol	8	1.0
Varaqua 936	120	13.0

PVC (%)	70.9
Total N.V. by Wt. (%)	46.7
Wt./gal (lb)	11.64
Visc. (KU)	85–90
Brushing	Excellent
Leveling	Very Good
Sagging Resistance	Excellent

Settling		None
Color Uniformity		Excellent
Recoat Uniformity		Excellent
Sheen Uniformity		Very Good

No. 6

Hot Pink

	lb	gal
Cellosize QP-4400 (2% Sol'n.)	175	21.00
Water	50	6.00
Ethylene Glycol	25	2.75
Nopco NDW	2	.25
Tamol 731 (25%)	10	1.00
Day-Glo Aurora Pink T-11	102	8.95
Titanium Dioxide R-900	100	2.90
Atomite	230	10.19

Procedure:

Add pigment slowly under good agitation using order of addition as listed. The Cowles Dissolver or equivalent equipment is recommended. Disperse until a paste temperature of 110–120 F is obtained.

Cool grind paste to room temperature and add the following under gentle agitation:

Nopco NDW	1	.12
Flexbond 315	400	44.0
(2% QP-4400 Sol'n.)	36	4.28
Yield (gal)		101
Pigment (%)		38.2
Total Solids (%)		57.6
PVC (%)		50.0
Wt./gal		11.10
Grind		6H
Gloss		Flat
Visc. (KU)		80–85

No. 7

(Hot Pink)

	lb	gal
Cellosize QP-4400 (2% Sol'n.)	132	15.84
Ethylene Glycol	25	2.75
Nopco NDW	2	.25
Tamol 731 (25%)	9	.90
Aurora Pink Dispersion WT-11	185	19.90
Titanium Dioxide **R-900**	100	2.90
Atomite	230	10.19

Procedure:

Add ingredients in order listed under good agitation using the Cowles Dissolver or equivalent equipment. Disperse until a paste temperature of 110–120 F is obtained. Cool grind paste to room temperature and add the following under gentle agitation:

	lb	gal
Flexbond 315	400	44.0
Water	35	4.20
Yield (gal)		101
Pigment (%)		38.4
Total Solids (%)		58.13
PVC (%)		50.0
Wt./gal (%)		11.0
Grind		6H
Gloss		Flat
Visc. (KU)		80–85

No. 8

(Vibrant Yellow)

	lb	gal
Cellosize QP-4400 (2% Sol'n.)	175	21.0
Water	50	6.0
Ethylene Glycol	25	2.75
Nopco NDW	2	.25
Tamol 731	5	.50
Day-Glo Saturn Yellow T-17 Pigment	85	7.44

	lb	gal
Titanium Dioxide **R-900**	65	1.88
Atomite	331	14.66

Procedure:

Add materials slowly under good agitation using the order of addition listed. The Cowles Dissolver or equivalent equipment is recommended. Disperse until a paste temperature of 110–120 F is obtained. Cool grind paste to room temperature and add the following under gentle agitation.

Nopco NDW	1	.10
Flexbond 315	287	31.50
Cellosize QP-4400 (2% Sol'n.)	125	15.00
Yield (gal)		101
Pigment (%)		45.8
Total Solids (%)		60.8
PVC (%)		60.0
Wt./gal (lb)		10.40
Grind		6H
Gloss		Flat
Visc. (KU)		85–90

No. 9

(Golden Goldenrod)

	lb	gal
Cellosize QP-4400 (2% Sol'n.)	175	21.0
Water	50	6.0
Ethylene Glycol	25	2.75
Nopco NDW	2	.25
Tamol 731	5	.50
Day-Glo Arc Yellow T16 Pigment	85	7.44
Titanium Dioxide **R-900**	65	1.88
Atomite	331	14.66

Procedure:

Add materials slowly under good agitation using the order of addition as listed. The Cowles Dissolver or equivalent equipment is recommended.

Disperse until a paste temperature of 110–120 F is obtained. Cool grind paste to room temperature and add the following under gentle agitation.

Nopco NDW	1	.125
Flexbond 315	287	31.50
Cellosize QP-4400 (2% Sol'n.)	125	15.00
Cal Ink 877-000-2503	23	2.4
Yield (gal)		103
Pigment (%)		41.6
Total Solids (%)		55
PVC (%)		60.5
Wt./gal (lb)		11.40
Grind		6H

No. 10

(Red)

It may be desirable for convenience of manufacturing or complete elimination of "dusting" to incorporate Day-Glo T Series pigments as a 55% solids dispersion in water. These Day-Glo water dispersions are available for all the standard T-Series colors. The following formula shows the use of such a dispersion and added materials to duplicate the preceding latex flat formula. Constants are the same as shown above.

	lb	gal
QP-4400 (2% Sol'n.)	58	7.0
Ethylene Glycol	25	2.75
Nopco NDW	1	.12
Tamol 731	2	.25
Day-Glo T-Series Water Dispersion	470	47.90
Flexbond 315	400	44.0
C.I. 877-0018	13	.77
C.I. 877-2501	3	.31

Fluorescent Vinyl Colors

lb/100 gal

FORMULA	No. 1 (Pink)	No. 2 (Red)	No. 3 (Red Orange)	No. 4 (Orange)	No. 5 (Orange Yellow)	No. 6 (Yellow)	No. 7 (Green)	No. 8 (Blue)
QP-4400 (2%)(1)	175	175	175	175	175	175	175	175
Water	50	50	50	50	50	50	50	50
Ethylene Glycol	25	25	25	25	25	25	25	25
Nopco NDW (2)	3	3	3	3	3	3	3	3
Tamol 731 (3)	5	5	5	5	5	5	5	5
T-Series Pigment	259-T-11	259-T-13	259-T-14	259-T-15	259-T-16	259-T-17	259-T-18	259-T-19
Let Down:								
Flexbond 315 (4)	400	400	400	400	400	400	400	400
Water	25	31	31	32	28	22	33	32
C.I. 877-0018 (5)	29	13	15	13	20	32	13	15
C.I. 877-2501 (6)	–	3	2	2	2	2	–	–

Procedure:

Add pigment slowly under good agitation using order of addition indicated above (numbers in parentheses). The Cowles dissolver or equivalent equipment is recommended. Grind until a paste temperature of 110–120 F is obtained. Cool the grind paste to room temperature and add letdown ingredients.

Exterior-Interior Vinyl Fluorescent, Red

	lb	gal
QP-4400 (2% Sol'n.)	69	8.32
Ethylene Glycol	25	2.75
Nopco NDW	1	.12
Tamol 731 (25%)	2	.25
Day-Glo T-Series Water Dispersion WT-13	470	47.90
Flexbond 315	400	44.00

Interior Vinyl Fluorescent Flat, Red

	lb	gal
QP-4400 (2% Sol'n.)	58	7.00
Ethylene Glycol	25	2.75
Nopco NDW	1	.12
Tamol 731 (25%)	2	.25
Day-Glo T-Series Water Dispersion WT-13	470	47.90
Flexbond 315	400	44.00
Titanium Dioxide Dispersion 866-0018	13	.77
Permanent Yellow Dispersion 877-2501	3	.31

Yield (gal)	102
Total Solids (%)	49.5
PVC	50.0
Wt./gal (lb)	9.40
Visc. (KU)	75–85
Grind	6H
Gloss	Flat

Alkyd Enamels

FORMULA	No. 1 (White, Semigloss)		No. 2 (White, Semigloss)	
	lb	gal	lb	gal
TiPure R-900	250	7.3	250	7.3
Kadox 515	15	0.3	15	0.3
Atomite	200	8.9	275	12.2
Cargill BB-6	500	63.0	—	—
Aroplaz 1266	—	—	466	58.0
Glyceryl-Mono-Oleate	8	1.0	8	1.0
Cobalt Naphthenate (6%)	4	0.5	3.6	0.4
Calcium Naphthenate (4%)	1.5	0.2	1.3	0.2
Antiskinning Agent	1.0	0.1	1.0	0.1
Mineral Spirits	122	18.7	134	20.5
PVC (%)	29.2		35.0	
Visc. (KU)	85–90		85–90	

No. 3

(Eggshell White)

	lb/100 gal
Titanium-Calcium Pigment (50% TiO_2)	500
Antisettling Agent	
Thixcin R or equivalent	3
Celite 499	90
Long-Oil Soya Alkyd (65% N.V.*)	427
Mineral Spirits	170
Cobalt Drier (6%)	2.4
Lead Drier (24%)	3.6
Antiskinning Agent (**Exkin No. 2** or equivalent)	2.0
lb/100 gal	90
Hegman Fineness	5½
60° Gloss	16
85° Sheen	25
Touch-Up Uniformity	Good

*Syntex 70, Lankyd 1407, or equivalent.

Procedure:

Disperse the pigment in part of the vehicle with a high-speed disperser, then add the remainder and the flatting agent listed above.

No. 4

(White)

	lb	gal
Roller mill grind:		
Rutile Titanium Dioxide	320.0	9.9
Amberlac 292X (50% in xylol)	172.6	21.6
Mix with:		
Amberlac 292X (50% in xylol)	469.0	58.7
Xylol	70.9	9.8
Cobalt Naphthenate (6%)	0.5	0.07

Wt./gal (lb)	10.3
Total Solids (%)	62.0
Pigment (%)	50
Vehicle (%)	50

Drier (metal on alkyd solids) 0.01% cobalt. Reduce to spray viscosity (approximately 50% solids) with xylol.

No. 5

(Flat)

	lb	gal
Titanium Dioxide	185	5.2
Calcium Carbonate **Vicron 15-15**	250	11.1
Talc	125	5.5
Celite 281	25	1.3
Bentone 38	2.5	.1
Low Odor Mineral Spirits	74	11.0
Soya Lecithin	2.5	.3
Varkyd 1532-30 LO	250	35.0
Roller mill:		
Varkyd 1532-30 LO	138	19.0

	lb	gal
Low Odor Mineral Spirits	80	12.0
Cobalt Naphthenate (6%)	1.5	.2
Lead Naphthenate (24%)	2	.2
Volatile ASA	1	.1

PVC (%)	66.0
Vehicle Nonvolatile (Wt.)	22.0
Total Nonvolatile by Wt. (%)	62.5
Wt./gal (lb)	11.20
Visc. (KU)	85–95
Brushing	Very good
Leveling	Very good
Sagging resistance	Excellent
Settling	None
Skinning	None
Gloss	Flat
Dry — set to touch	0.5–1.0 h
hard	2–3
Color properties	Very good
Sheen uniformity	Very good

No. 6

(Gloss – Tint Base)

	lb	gal
Varkyd 553-50	145	19.0
Mineral Spirits	26	4.0
Thixatrol ST	4	.5
Titanium Dioxide **Ti-Pure R-900**	125	3.6
Calcium Carbonate **Atomite**	150	6.7
Soya Lecithin	2	.3

Intensive mixing to 140 F and mill

	lb	gal
Varkyd 553-50	395	52.0
Mineral Spirits	104	16.0
Cobalt Naphthenate (6%)	2	.3
Calcium Naphthenate (4%)	3	.4
Lead Naphthenate (24%)	5	.5
Vol. Antiskinning Agent **Exkin #2**	1	.1

	lb	gal
Advance Puffing Agent	7.6	1.0
Blown Soya Oil Z_3	8	1.0

PVC (%)	24.5
Vehicle Nonvolatile (% Wt.)	41.3
Total N.V. by Wt. (%)	57.5
Wt./gal (lb)	9.26
Visc. (KU)	75–80
Brushing	Excellent
Leveling	Excellent
Sagging Resistance	Excellent
Settling	None
Skinning	None
Gloss	Excellent
Dry — Set-to-Touch	1–2 hr
— Hard	4–6 hr

No. 7

(Gloss)

	lb	gal
Rutile Titanium Dioxide	344.00	9.85
Thixatrol ST	7.70	—
Plaskon ST-860 (60% NV)	573.00	71.70
Mineral Spirits	143.00	21.70
Cobalt Naphthenate (6%)	2.30	—
Calcium Naphthenate (4%)	5.10	—
Exkin #2	3.40	—

Disperse pigment and **Thixatrol ST** with portion of **Plaskon ST-860** using suitable equipment to achieve minimum temperature of 140°F. Reduce pigment paste with balance of formula.

Pigment by Wt. (%)	32.1
Resin Solids by Wt. (%)	31.9
Total Solids by Wt. (%)	64.0
PVC (%)	21.1
Visc. at 25 C (KU)	89

No. 8

(Gloss)

	lb	gal
Titanox C-50	258.0	8.92
Titanium Dioxide **R-900**	85.0	2.49
Nev Chem 120-70	83.0	10.35
Plaskon-3139	171.0	21.50
Plaskon-633	13.8	1.87
Plaskon-3893	5.0	0.57
Thixcin GR	5.0	—
Troykyd Antifloat	2.0	—
Maglite D Dispersion	16.5	1.52

Grind above on roller mill, then add:

	lb	gal
Plaskon-3139	180.0	22.70
Amsco 365	179.0	26.90
Amsco 140H	47.0	6.90
Cobalt Naphthenate (6%)	1.8	—
Calcium Naphthenate (4%)	1.8	—
Exkin #2	0.8	—

Visc. 25 C KU (overnight)	90
Pigment (by Wt.) (%)	32.9
Resin Solids (by Wt.) (%)	29.8
Total Solids (by Wt.) (%)	62.7

No. 9

(White Silicone)

	lb	gal
Nonchalking Titanium Dioxide	275	7.85
Soya Lecithin	4	0.50
Varkyd 385-50X	616	74.20
Bentone 38	5	0.40
S. Cal #3	52	7.0
Xylene	72	10.0
Cobalt Naphthenate (6%)	4.0	0.35
Lead Naphthenate (24%)	6.0	0.75
Exkin #2	1.0	0.15

PVC (%)	20
Vehicle Nonvolatile (% Wt.)	41.6
Total NV by Wt.	57.2
Wt./gal (lb)	10.2
Visc. (KU)	70–80
Reduction: 15% Xylene	
Reduction visc. #4 Ford	20–25

Epoxy Enamel

FORMULA NO. 1

(White)

		lb
A	**Araldite 7072**	300.0
	Rutile Titanium Dioxide	245.0
	Xylene	164.0
	Cellosolve	63.0
	MIBK	63.0
	Flow Control Agent	2.0
	Antisag Agent	3.0
B	**Araldite Hardener 835**	150.0
	Solvesso 150	30.0

Performance:

Cure Schedule	10 days at 25°C (77°F) and 50% RH
Substrate	Steel
Film Thickness, mil	4
Flexibility, 1/8 inch mandrel	Pass
Adhesion	Excellent
Gloss, 60 Geometry	98
Impact Resistance, in. lb	
Pass, direct	140
Impact Resistance, in. lb	
Pass, reverse	160
Resistance at 25 C (77 F) to	
water, 7 days	Unaffected
MIBK, 3 days	Sl. softening
10% HCl 8 h	Unaffected

Resistance to salt fog, 200 h	Unaffected
Humidity, 10 days	Blisters, No. 8 dense
Weatherometer, 250 h	Sl. yellow discoloration and slight loss of gloss

No. 2

(White Gloss)

		lb
A	**Araldite 540 X-90**	371.2
	Titanium Dioxide	475.2
	Xylene	93.3
	n-Butanol	47.7
	MIBK	19.9
	Flow Control Agent	18.2
B	**CIBA Polyamide 825**	120.6
	Xylene	35.8

No. 3

(White Gloss)

		lb
A	**Araldite 571 KX-75**	308.0
	Rutile Titanium Dioxide	232.2
	MIBK	93.1
	Methyl **Cellosolve**	131.7
	Xylene	39.8
	Antisag Agent	4.7
B	**CIBA Polyamide 815 X-70**	177.7

Nonvolatile Content (%)	60.0
Epoxy/Hardener Ratio	100/54
Pigment/Binder Ratio	40/60
Density, lb/gal 25°C (77°F)	9.9
Pot Life, h, min, 25°C (77°F), 1 lb	>48.00
Visc., 25°C (77°F) (Parts A & B), KU	64

No. 4

(White Gloss)

		lb
A	**Araldite 7072**	166.0
	Titanium Dioxide	308.0
	Xylene	117.6
	Cellosolve	78.4
	Antisag Agent	2.0
B	**CIBA Polyamide 800 CX-60**	256.0
	Xylene	54.0
	Cellosolve	36.0

No. 5

(White Gloss)

		lb
A	**Araldite 571-T75**	309.0
	Rutile Titanium Dioxide	281.0
	Calcium Carbonate	77.0
	Cellosolve	34.0
	Flow Control Agent	2.0
	Antisag Agent	3.0
B	**CIBA Polyamide 815 X-70**	177.0
	Toluene	121.0
	Cellosolve	13.0
	Solvesso 150	43.0
	MEK	21.0
Nonvolatile Content (%)		66.0
Epoxy/Hardener Ratio		100/54
Pigment/Binder Ratio		50/50
Density lb/gal, $25°C$ ($77°F$)		10.8
Visc., $25°C$ ($77°F$) KU		70

No. 6

(White)

		lb	gal
A	**Resypox 1571**	196.1	19.52
	MIBK	64.4	9.63
	Cellosolve Solvent	46.6	6.01
	Xylene	56.8	7.87
	Cyclohexanol	7.6	0.95
	Beetle 216–8	12.8	1.43
B	**Resycure 313**	104.3	12.56
	MIBK	94.1	14.10
	Butyl Cellosolve	10.5	1.40
	Toluene	106.0	14.63
	Denatured Ethyl Alcohol	24.8	3.66

Pigments:

	lb	gal
Ti Pure R610	240.2	6.87
Phthalocyanine Blue	Trace	Trace
Asbestine 3–X	32.5	1.37

Procedure:

Prepare 60% NV Solution of **Resypox 1571** and grind in pigments. Let down with remaining solvent and add **Beetle 216–8**. Mix thoroughly and add curing agent solution.

No. 7

(Gray)

A	**Resypox 1571-T**	31.8
	Titanium Dioxide	12.7
	MIBK	8.0
	Beetle 216-8	0.4
	Xylene	2.8
	Yellow Iron Oxide	2.2
	Carbon Black	1.0
	Tinting Pigment	Trace
B	**Resymide 1415**	18.6
	Xylene	15.0

Butyl Alcohol	4.3
Isopropyl Alcohol	3.2

Polyvinyl Acrylic Latex Enamel

FORMULA NO. 1

(Hot Pink, Satin)

	lb	gal
Aurora Pink Dispersion WT-11	182	19.60
Tamol 731 (25%)	3	0.30
Drew L-475	4	0.50
R-900 Titanium Dioxide	100	2.90

Procedure:

Add ingredients in order shown under good agitation using the Cowles Dissolver or equivalent equipment. Disperse until a paste temperature of 110–130 F is obtained.

Cool grind paste to room temperature and add the following under gentle agitation:

Propylene Glycol	77	8.93
Ethylene Glycol	18	2.00
Wallpol 40-134	602	67.30
Drew L-475	2	0.25
Yield (gal)	102	
Pigment (%)	20.0	
Total Solids (%)	51.31	
PVC (%)	26.0	
Wt./gal	9.69	
Grind	6H	
Gloss	15:60° Head	
Visc. (KU)	70–75	

No. 2

(Hot Pink, Satin)

	lb	gal
Propylene Glycol	77.0	8.93
Ethylene Glycol	30.0	3.24
Cellosize QP-4400 (2% Sol'n.)	36.0	4.40
Tamol 731 (25%)	4.5	0.50
Drew L-475	4.0	0.50
Day-Glo Aurora Pink T-11	100.0	8.75
Titanium Dioxide R-900	100.0	2.90

Procedure:

Add pigment slowly under good agitation using order of addition indicated above. The Cowles Dissolver or equivalent equipment is recommended. Disperse until a paste temperature of 110–120 F is obtained. Cool Grind paste to room temperature and add the following under gentle agitation:

Drew L-475	4.0	.50
Wallpol 40-134	602.0	67.30
Cellosize QP-4400 (2% Sol'n.)	42.0	5.00

Yield (gal)	102
Pigment (%)	20.0
Total Solids (%)	50.75
PVC (%)	26.6
Wt./gal (lb)	9.79
Grind	6H
Gloss	15:60° Head
Visc. (KU)	70–75

No. 3

(Golden Goldenrod)

	lb	gal
Propylene Glycol	77	8.93
Ethylene Glycol	30	3.24
Cellosize, QP-4400 (2% Sol'n.)	36	4.40

	lb	gal
Tamol 731 (25%)	5	.50
Drew L-475	4	.50
Day-Glo Arc Yellow T-16 Pigment	85	7.44
Titanium Dioxide **R-900**	65	1.88

Procedure:

Add materials slowly under good agitation using order of addition indicated above. The Cowles Dissolver or equivalent equipment is recommended. Disperse until a paste temperature of 110–120 F is obtained. Cool grind paste to room temperature and add the following under gentle agitation.

Drew L-475	4	0.50
Wallpol 40-134	602	67.30
Cellosize QP-4400 (2% Sol'n.)	58	7.0
Cal Ink 877-000-2503	19	2.02

Yield (gal)	104
Pigment (%)	15.8
Total Solids (%)	47
PVC (%)	23.4
Wt./gal (lb)	9.47
Grind	6H
Gloss	15–20:60° Head
Visc. (KU)	70–75

No. 4

(Vibrant Yellow)

	lb	gal
Propylene Glycol	77	8.93
Ethylene Glycol	30	3.24
Cellosize QP-4400 (2% Sol'n.)	36	4.40
Tamol 731	4	0.50
Drew L-475	4	0.50
Day-Glo Saturn Yellow T-17 Pigment	85	7.44
Titanium Dioxide **R-900**	65	1.88

Procedure:

Add materials slowly under good agitation using order of addition indicated above. The Cowles Dissolver or equivalent equipment is recommended. Disperse until a paste temperature of 110--120 F is obtained. Cool grind paste to room temperature and add the following under gentle agitation.

Drew L-475	4	0.50
Wallpol 40-124	602	67.30
Cellosize QP-4400 (2% Sol'n.)	58	7.00

Yield (gal)	102
Pigment (%)	15.50
Total Solids (%)	47.30
PVC (%)	22.40
Wt./gal (lb)	9.46
Grind	6H

Acrylic Enamel

FORMULA No. 1

(Semigloss)

	lb/100 gal	kg/100 ℓ
Grind:		
Pigment Dispersant (25%)	11	1.320
Defoamer	2	0.240
Water	73	8.76
Propylene Glycol	32	3.84
Titanium Dioxide	275	33
Let Down:		
Dowicil 75 Preservative	1.33	0.160
Water	20	2.4
Propylene Glycol	32	3.84
Dalpad A Coalescing Agent	10	1.2
Defoamer	4	0.480
Surfactant	2	0.240
Dowfax 2A1 Surfactant	2	0.240

	lb/100 gal	kg/100 ℓ
Acrylate Copolymer Acrylic		
Latex (46.5%)	590	70.8
Water	70	8.4
Methocel J12MS	3–4	0.360–0.480
PVC (%)		19.7
Nonvolatile Content (%)		48.7
pH		9–10
Visc. (KU)		80–90

Other surface treated **Methocel** products (**J5MS, J20MS, J75MS, K4MS, K15MS**) may be substituted.

	No. 2		No. 3		No. 4		No. 5	
	(Semigloss)				*(Egg Shell)*			
	(White)		*(Tint Base)*		*(White)*		*(Tint Base)*	
	lb	gal	lb	gal	lb	gal	lb	gal
1. **Ti-Pure R-900**	250	7.3	150	4.4	250	7.3	150	4.4
2. **Atomite**	25	1.1	25	1.1	125	5.5	100	4.4
3. **Daxad 30**	5	0.5	5	0.5	5	0.5	5	0.5
4. **Colloid 606**	2	0.3	2	0.3	2	0.3	2	0.3
5. Ethylene Glycol	28	3.0	28	3.0	28	3.0	28	3.0
6. Butyl Cell. Acetate	7.8	1.0	7.8	1.0	7.8	1.0	7.8	1.0
7. **Troysan PMA-30**	0.3	–	0.3	–	0.3	–	0.3	–
8. **Igepal CO-610**	2	0.2	2	0.2	2	0.2	2	0.2
9. **Natrosol 250 H.R.** (3%)	8	1.0	75	9.0	8	1.0	92	11.0
10. Water	88	10.6	196	23.5	127	15.2	202	24.2

	No. 2		No. 3		No. 4		No. 5	
	(Semigloss)				*(Egg Shell)*			
	(White)		*(Tint Base)*		*(White)*		*(Tint Base)*	
	lb	gal	lb	gal	lb	gal	lb	gal
11. Rhoplex AC-490	655	75.0	499	57.0	577	66.0	446	51.0
PVC (%)		21.3		18.8		31.9		29.3
Visc. (KU)			75–80				75–80	
Initial Gloss (60° Gloss Meter)			40–50				15–25	

Procedure:

Add the following ingredients in the order given, with good mixing: Water (as necessary to obtain reasonable mixing consistency), **Daxad 30**, prime pigment and extender. Mix at high speed until pigment slurry is completely dispersed and homogeneous (about 30 min or longer depending on efficiency of mixer), *then reduce speed of mixer*. Add the remaining water, followed by ingredients 4, 5, 6, 7, and 9. Mix thoroughly after each addition. Add the latex binder and mix for about 5 min. Finally, add the surfactant (**Igepal CO-610**); mix thoroughly and then package.

No. 6

(White Spray)

The following formulation based on **Acryloid B-67** as a 45% solids solution in VM & P naphtha is a general purpose white spray enamel.

	lb	gal
Roller mill grind:		
Rutile Titanium Dioxide	153.2	4.39
Acryloid B-67 (45%)	90.8	12.62
VM & P Naphtha	11.2	1.79
Mix with:		
Acryloid B-67 (45%)	419.9	58.30

	lb	gal
VM & P Naphtha	52.5	8.36
Mineral Thinner	94.5	14.54
Wt./gal (lb)		8.2
Total Solids (%)		46.6
Titanium Dioxide		40%
Acryloid B-67		60%

No. 7

(Spray)

	lb	gal
Roller mill grind:		
Rutile Titanium Dioxide	277.0	7.95
Zinc Oxide	14.6	0.29
Duraplex D-65A (70%)	194.4	24.63
Mix with:		
Duraplex D-65A (70%)	118.1	14.96
Acryloid B-67 (45%)	162.0	22.53
VM & P Naphtha	20.3	3.21
Mineral Thinner	170.1	26.20
Cobalt Naphthenate (6% metal)	1.8	0.23
Wt./gal (lb)		9.6
Total Solids (%)		60.8
Pigment (%)		50
Vehicle (%)		50
Acryloid B-67		25%
Duraplex D-65A		75%
Visc. (No. 4 Ford Cup)		73 s

Enamel should be made to an approximate viscosity of 110 s then allowed to equilibrate for about 3 days and adjusted to 73 s.

Film Properties of Modified Enamel, Air-Dried:

Set time	½ h
Tack-Free Time (500 gw)	5 h

Hardness (Tukon), 7 days	2.8
Adhesion (H value)	15.6
Photovolt Gloss	86
Whiteness (Photovolt K)	9.6
Yellowing (Photovolt K)	
1 month (dark)	13.2
115 h at 140°F. (dark)	13.9

Duraplex Air-Drying Enamel

Formula No. 1

(Chlorinated Rubber)

	lb
Parlon 10 cp (22.1% solids in hi-flash naphtha)	60.5
Duraplex C-55X (70% solids)	37.8
Aluminum Paste	2.0
Butanol	2.0
Cobalt Naphthenate Drier (6%)	0.24

	Solids Basis
Parlon	32.1%
Duraplex C-55X	63.4%
Aluminum Powder	4.5%
Total Solids (%)	41.0%

No. 2

(Red)

	lb	gal
Roller mill grind:		
Toluidine Red	56.1	4.6
Duraplex C-55 (50% solids)	112.2	14.8
Mineral Thinner	18.7	2.9
Mix with:		
Duraplex C-55 (50% solids)	336.2	44.2
Mineral Thinner	102.8	15.8
V.M.&P. Naphtha	102.8	16.8

	lb	gal
Cobalt Naphthenate (6%)	2.8	0.4
Calcium Naphthenate (6%)	3.7	0.5

Wt./gal (lb)	7.4
Total Solids (%)	39
Pigment (%)	20
Binder (%)	80

Drier (Metal based on alkyd solids)

Cobalt	0.075%
Calcium	0.10%

For spray application this enamel should be reduced with xylol or **Amsco B** to 21 s on a Ford #4 Cup.

White Gloss Topcoat
Enamel

(Polyvinyl Chloride)

	lb	gal
Vinoflex MP 400 (Medium)	135.0	13.1
Titanox CL	250.0	7.2
Thixatrol MPA 60 (Xylene)	75.0	10.3
Xylene	300.0	41.7
Mineral Spirits	180.0	27.7

Procedure:

Ground 16 h in a high density alumina jar mill.

PVC (%)	29.0
Nonvolatile (%)	45.7
Nonvolatile by vol. (%)	30.6
Wt./gal (lb)	9.40
Visc. (KU)	69
Visc., S DIN #4 Cup	35
(20% Reduction by volume with Toluene)	

Tint Bases

A tint base is a white paint to which a color pigment is added to produce a pastel paint. The final color produced is composed of 50% or more of the white pigment.

Latex Paint-Tint Base

FORMULA No. 1

(PVA)

	lb	gal
Water	150	18.0
Tamol 731-25	7	.8
Foamicide KP-140	3	.4
Ethylene Glycol	20	2.1
Amsco Odorless Mineral Spirits	3.1	.5
Texanol Solvent	8	1.0
Whiting Snowflake	150	6.7
Microcel T-38 J.M.	25	1.3
Titanium Dioxide Ti-Pure R-901	200	6.0
Cellosize QP-15,000 (2½% Sol'n.)	100	12.0
Talc Fibrene C-500	100	4.4
Mildewcide Metasol 57 Liquid	.5	—

Mill and thin down

	lb	gal
Cellosize QP-15,000 (2½% Sol'n.)	100	12.0
Varaqua 928	264	29.0
Water	75	9.0

PVC (%)	53.8
Vehicle Nonvolatile (% Wt.)	21.1
Total N.V. by Wt. (%)	52.2
Wt./gal (lb)	11.69
Visc. (KU)	90–95
Application Properties	Excellent
Color Uniformity	Excellent
Recoat Uniformity	Excellent
Sheen Uniformity	Very Good
Scrub Resistance	Very Good

No. 2

(White and Light)

	lb	gal
Pigment Grind:		
Water	275.0	33.0
Advawet 33	4.0	0.4
Colloid 677	2.0	0.2
Tamol 731	6.0	0.6
Potassium Tripolyphosphate	1.0	0.1
Dowicil 75	1.0	–
Ethylene Glycol	30.0	3.2
Cellosize Hydroxyethyl **Cellulose**		
QP-4400H	4.5	0.4
Ti-Pure R-901	275.0	8.4
Micro-Cel T-70	30.0	1.6
Gold Bond R	50.0	2.3
Optiwhite	50.0	3.2
Ucar Filmer 351	10.0	1.3
Let down:		
Ucar Latex 365	255.0	28.2
Colloid 677	1.0	0.1
Water	142.0	17.0
PVC (%)		51.5
Total Solids (%)		49.0
Angular Sheen (85°)		2.5
Contrast Ratio		0.98
Visc. (KU)		85–90
Wt./gal (lb)		11.3–11.4
Brightness (%)		96

No. 3

(White and Light)

	lb	gal
Ethylene Glycol	10.0	1.1
Water	200.0	24.0

	lb	gal
CMP Acetate	2.5	0.3
KTPP	5.0	0.2
Nopcosant K	10.0	0.9
Ammonium Hydroxide (28%)	2.0	0.3
Natrosol 250 MR (3% sol'n.)	50.0	6.0
Foamaster VL	2.0	0.26
Ti-Pure R-901	115.0	3.29
Icecap K	192.0	8.95
Optiwhite	125.0	6.81
Gold Bond R	50.0	2.28

Grind with a Cowles type mixer for 20 min (4800 fpm). Add the following at lower speed:

	lb	gal
Water	41.0	5.0
Polyco 2407	240.0	28.4
Natrosol 250 MR (3% sol'n.)	100.0	12.0
Foamaster VL	2.0	0.26
PVC (%)	61.5	
Total Solids (%)	52.6	
Wt./gal (lb)	11.5	

Acrylic Copolymer Interior Tint Base

	lb	gal
Premix:		
Water	200.0	24.00
AMP-95	4.0	0.51
Disperse pigments:		
Horse Head R-770 Titanium Dioxide	150.0	4.60
Snowflake Whiting	200.0	8.86
ASP 400 Clay	125.0	5.83
Grind; then add:		
Ethylene Glycol } Premix	30.0	3.24
Preservative }	0.5	0.05

	lb	gal
Cellosize QP-4400 (2½% Thickener)	232.0	27.86
Butyl Carbȋtol	15.0	1.90
Ucar Latex 380 Acrylic Copolymer Emulsion (49%)	200.0	22.67
Defoamer	2.0	0.26
Triton CP-10 Surfactant	2.0	0.22

Visc. (KU) at 77°F — Fresh	92
Overnight	95
1 month	96
1 month at 125°F	99
pH — Fresh	9.2
1 month at 125°F	8.7
PVC (%)	61.3
Total Solids (%)	49.5

Vinyl Acrylic Flat — Tint Base

Formula No. 1

	lb	gal
Water	150	18.0
Tamol 731-25	9	1.0
Tritonex 102	4	.5
Propylene Glycol	26	3.0
Foamicide NDW Nopco	2	.3
Titanium Dioxide **Ti-Pure R-901**	200	6.0
Whiting **Snowflake**	200	8.9
Cellosize QP-15,000 (2.5% Sol'n.)	100	12.0
Talc **Chemet Ruby 400**	100	4.4
Mildewcide **Metasol 57** Sol'n.	0.5	—
Cellosize QP-15,000 (2.5% Sol'n.)	125	15.0
Texanol	16	2.0
Water	75	9.0
Varaqua 936	209	23.0

PVC (%)	59.8
Total N.V. by Wt. (%)	51.5
Wt./gal (lb)	11.78
Brushing	Excellent

Leveling	Very Good
Sagging Resistance	Excellent
Gloss	Flat
Color Uniformity	Excellent
Recoat Uniformity	Excellent
Sheen Uniformity	Very Good +
Visc. (KU)	75–80

No. 2

	lb	gal
Water	150	18.9
Tamol 731-25	9	1.0
Tritonex 102	4	.5
Propylene Glycol	26	3.0
Foamicide **Troykyd 666**	3	.4
Titanium Dioxide **Ti-Pure R-901**	200	6.0
Whiting **Snowflake**	200	8.9
Cellosize QP-15,000 (2.5% Sol'n.)	100	12.0
Talc **Chemet Ruby 400**	100	4.4
Microcel T-38	25	1.3
Mildewcide **Metasol 57** Sol'n.	0.5	—
Cellosize QP-15,000 (2.5% Sol'n.)	100	12.0
Texanol	16	2.0
Water	75	9.0
Varaqua 936	236	26.0

PVC (%)	58.8
Total N.V. by Wt. (%)	53.3
Wt./gal (lb)	11.92
Visc. (KU)	90–95
Brushing	Excellent –
Leveling	Very Good
Sagging Resistance	Excellent
Settling	None
Gloss	Flat
Scrub Resistance (cycles) 3 mils	
Sandblased Glass (2.5% Ivory Soap)	Average 2686
Color Uniformity	Excellent
Recoat Uniformity	Excellent –
Sheen Uniformity	Very Good

No. 3

(Light and White)

	lb	gal
Water	175	21.0
Tamol 731-25	10	1.1
Triton X-102	4	.5
Propylene Glycol	26	3.0
Foamicide **Troykyd 666**	2.5	.3
Titanium Dioxide **Ti-Pure R-901**	275	8.25
Whiting **Snowflake**	150	6.65
Talc **Chemet's Ruby 400**	100	12.0
Cellosize QP-15,000 (2½% Sol'n.)	100	4.4
Microcel T-38	25	1.3
Mildewcide **Metasol 57** Sol'n.	0.5	—
Cellosize QP-15,000 (2.5% Sol'n.)	100	12.0
Texanol	16	2.0
Water	50	6.0
Varaqua 936	236	26.0

PVC (%)	58.8
Total N.V. by Wt. (%)	54.5
Wt./gal (%)	12.14
Visc. (KU)	95–100
Brushing	Excellent
Leveling	Very Good
Sagging Resistance	Excellent
Settling	None
Gloss	Flat
Color Uniformity	Excellent
Recoat Uniformity	Excellent
Sheen	Very Good

Primers and Primer-Sealers

Though the primer is not the final coat, it is the key to the success of the paint system. An unsuccessful primer causes metal to corrode, wood to rot, and blistering and flaking to occur. The primer or primer sealer must act as a bridge between the surface it is covering and the topcoat.

The important characteristics of a good primer or primer-sealer paint are: good flow and leveling, complete enamel holdout, wet-rub resistance, good intercoat adhesion, resistance to lime burn, fast drying, and minimum cost.

Alkyd Enamel Primer

(Odorless)

	lb	gal
Titanox RCHT	500.0	18.45
Calcium Carbonate	50.0	2.21
Celite 281	20.0	1.04
High Acid Al. Stearate	2.0	.24
Odorless Flat Alkyd	310.0	43.00
Odorless Long Oil Alkyd	102.0	13.50
Lead Naphthenate (24%)	1.5	.16
Cobalt Naphthenate (6%)	.6	.08
Antiskinning Agent	1.0	.13
Odorless Mineral Spirits	139.1	21.19

Pigment (%)	50.8
Vehicle (%)	49.2
Vehicle N.V. (%)	33.3
PVC (%)	51.7

Polyurethane Primer

(White)

	lb	gal

Slurry Grind Method

Pebble mill charge:

	lb	gal
Rutile Titanium Dioxide	150.0	4.3
Extended Titanium Dioxide	100.0	3.6
Talc **(Nytal 300)**	50.0	2.1
Celite 281	50.0	2.6
Mica (325 Mesh Waterground)	50.0	2.1
Cellosolve Acetate	60.0	7.4
Xylol	310.0	43.0
Tolylene Diisocyanate **(Nacconate 80)**	10.5	1.1

	lb	gal
Tumble overnight, add		
Spenkel M86-50CX	330.0	38.4
Visc. (KU)	56	
PVC (%)	45	
Vechile solids (%)	27.3	

Latex Primer-Sealer

Formula No. 1

	lb	gal
Water	50.0	6.00
Ethylene Glycol	23.3	2.50
Texanol	4.8	0.60
Daxad 30	6.4	0.67
Triton X-100	3.0	0.34
Nopco NXZ	1.9	0.25
Super-Ad-It	0.5	0.06
Rutile Titanium Dioxide	75.0	2.15
Asbestine 325	100.0	4.22
Atomite	175.0	7.77
Celite 281	25.0	1.31
Natrosol 250 HR (2% Sol'n.)	225.0	27.00

Grind above in Cowles dissolver, then premix following and blend with **Poly-Tex 668** and dibutyl phthalate before mixing with grind. Use gentle agitation:

	lb	gal
Poly-Tex 668	288.0	31.00
Dibutyl Phthalate	5.0	0.57
Water	77.5	9.31
Nopco NXZ	1.9	0.25

Use following as needed for viscosity adjustment:

	lb	gal
Water (or Protective Colloid Sol'n.)	50.0	6.00
Visc. (KU)	94 Fresh, 95 overnight	

	lb	gal
Wt./gal		11.12
PVC (%)		45.2
Vehicle Solids (%)		25.2
Total Solids (%)		50.6
% Mercury		0.0045

No. 2

Grind:	lb	gal
Water	83.3	10.00
Dispersant, **Daxad 30**	5.6	.58
Coalescing Agent, **Dalpad A**	7.0	.76
Ethylene Glycol	20.0	2.16
Rutile Titanium Dioxide **Zopaque R 55**	100.0	2.96
Clay, **ASP-200**	125.0	5.82
Celite 281	25.0	1.30

Disperse using high speed disperser adding in above order.

Let Down:	lb	gal
Water	150.0	18.00
Preservative, **Dowicide O**	1.0	.10
Antifoam, **Colloids 677**	1.0	.15
Polyvinyl Acetate Emulsion, **Polyco 804**	364.6	40.00
Thickener (2% Sol'n.) **QP-4400**	150.0	18.00
Ammonia to give pH 7.5–8.5	as needed	
Wt./gal	10.3	
Nonvolatile (%)	46.7	
PVC	32.3	

Polyvinyl Acetate Primer and Sealer

Premix:	lb	gal
Ethylene Glycol	20.0	2.15
Water	150.0	18.00

	lb	gal
AMP-95	3.0	0.38
Disperse pigments:		
Ti-Pure R-931 Titanium Dioxide	100.0	3.16
Camel-Carb Dry Ground Whiting	100.0	4.43
Asbestine 3X Talc	100.0	4.20
Grind: then add		
Preservative	0.5	0.05
Cellosize QP-4400 (2½% Sol'n.) Thickener } Premix	208.5	25.03
Everflex BG Vinyl Acetate Copolymer Emulsion (52%)	350.0	39.38
Butyl Carbitol	27.0	2.84
Defoamer	2.0	0.26
Igepal CO-630 Surfactant	1.0	0.12

Visc. (KU) at 77°F – Fresh	84
Overnight	83
1 week	84
1 month	84
1 month at 125°F	90
pH – Fresh	8.8
1 month at 125°F	7.3
PVC (%)	38

Ceiling Paints

The major considerations in manufacturing ceiling paints are getting the most hiding power with a minimum of cost and still maintaining film integrity. Since paints used for this type of application are generally white or off-white and will not be washed or scrubbed with any degree of regularity, they can have a significantly lower film integrity than that required for a woodwork or wall paint.

Alkyd Ceiling Paint

FORMULA	No. 1		No. 2		No. 3	
	lb	gal	lb	gal	lb	gal
Titanox RCHT	600	22.1	—	—	—	—
Titanox C-50	—	—	360	12.5	—	—
Titanox RA-50	—	—	—	—	180	5.1
Atomite	—	—	130	5.8	230	10.2
Duramite	250	11.1	250	11.1	250	11.1
Al-Sil-Ate O	—	—	82	3.8	146	6.8
Thixcin	3	0.4	3	0.4	3	0.4
1952-40 Wallkyd (40 NV)	315	43.0	315	43.0	315	43.0
Cobalt Drier (6%)	1	0.1	1	0.1	1	0.1
Lead Drier (24%)	2	0.2	2	0.2	2	0.2
Antiskinning Agent	2	0.2	2	0.2	2	0.2
Low Odor Mineral Spirits	145	22.9	145	22.9	145	22.9
Visc. (KU)	80–90		80–90		80–90	
PVC (%)	71.0		71.0		71.0	
Contrast Ratio (%)	92.7		93.0		93.6	
Brightness (%)	88.6		88.5		88.0	

Paints made according to these formulations are quite similar in all respects except sheen. The lowest sheen results when pure titanium dioxide is used with the extender combination composed of **Atomite, Duramite,** and **Al-Sil-Ate O.** The paints made with 30% titanium-calcium pigment are significantly higher in sheen, and those made with the 50% titanium-calcium pigment plus the same three extenders are intermediate in sheen.

In order to make the three paints substantially equal in sheen also, it is necessary to replace some of the **Duramite** flatting pigment from the paint containing 50% titanium-calcium pigment and also from the pure rutile titanium dioxide containing formulation and replace this with an equal volume of **Atomite.** Since this move increases the total surface area of the pigmentation it is necessary to remove some of the calcined clay and replace it with an equal volume of **Atomite** such that the surface area equality is retained and also the same PVC is retained. The three formulations after balancing for equal hiding, sheen, and film integrity are given as follows:

	No. 4		No. 5		No. 6	
	lb	gal	lb	gal	lb	gal
Titanox RCHT	600	22.1	—	—	—	—
Titanox C-50	—	—	360	12.5	—	—
Titanox RA-50	—	—	—	—	180	5.1
Atomite	—	—	187	8.3	400	17.9
Duramite	250	11.1	200	8.9	100	4.4
Al-Sil-Ate O	—	—	75	3.5	125	5.8
Thixcin	3	0.4	3	0.4	3	0.4
1952-40 Wallkyd						
(40 NV)	315	43.0	315	43.0	315	43.0
Cobalt Drier (6%)	1	0.1	1	0.1	1	0.1
Lead Drier (24%)	2	0.2	2	0.2	2	0.2
Antiskinning Agent	2	0.2	2	0.2	2	0.2
Low Odor Mineral						
Spirits	145	22.9	145	22.9	145	22.9
Visc. (KU)	80–90		80–90		80–90	
PVC (%)	71.0		71.0		71.0	
Contrast Ratio (%)	92.7		93.5		94.9	
Brightness (%)	88.6		88.6		88.4	

	No. 7		No. 8		No. 9	
	lb	gal	lb	gal	lb	gal
Titanox RCHT	600	22.1	—	—	—	—
Titanox C-50	—	—	360	12.5	—	—
Titanox RA-50	—	—	—	—	180	5.1
Atomite	—	—	350	15.6	690	30.6
Duramite	250	11.1	200	8.9	125	5.7
Thixcin	3	0.4	3	0.4	3	0.4
1952-40 Wallkyd						
(40 NV)	315	43.0	286	39.2	240	32.8
Cobalt Drier (6%)	1	0.1	1	0.1	1	0.1
Lead Drier (24%)	2	0.2	2	0.2	2	0.2
Antiskinning Agent	2	0.2	2	0.2	2	0.2
Low Odor Mineral						
Spirits	145	22.9	145	22.9	157	24.9
Visc. (KU)	80–90		80–90		80–90	

PVC (%)	71.0	75.1	80.0
Contrast Ratio (%)	92.7	93.2	93.0
Brightness (%)	88.6	87.4	88.1

Latex Ceiling Paint

(Vinyl Chloride)

	lb	gal
Titanox RANC	100.0	2.8
ASP 400	185.5	8.6
Celite 281	100.0	5.2
Atomite	152.8	6.8
Potassium Tripolyphosphate	2.0	—
Tamol 731	2.0	.2
Water	222.1	26.6
Cellosize WP 4400 (2½% Sol'n.)	200.0	24.0
Texanol	3.3	.4
Ethylene Glycol	20.4	2.2
Geon 450X20	112.0	11.9
Colloids 60	2.0	.2
Potassium Carbonate	2.2	—
Water	113.6	13.6

PVC (%)	80.13
Wt./gal (lb)	11.88
Nonvolatile (%)	49.98
Nonvolatile vol. (%)	28.48
Visc. (KU)	74.00
pH	9.10
85° Sheen	0.00
Scrub cycles, w/Bon Ami	100

Acrylic Ceiling Paint

	lb	gal
Water	200.0	24.00
Tamol 731	8.0	1.00
Igepal CO-630	4.0	.50
TKPP	1.0	—

	lb	gal
Latekoll O (25%)	24.0	2.76
PMA-30	0.3	.03
Titanox RA-47	75.0	2.37
Micro-cel T-70	30.0	1.59
ASP-400	125.0	5.81
Drikalite	300.0	13.29
Crystalline Silica #219	50	2.27
Concentrated Ammonia	2.0	.25

Add in order listed and mix on a high speed disperser for 10 min, then let down as follows:

	lb	gal
Ethylene Glycol	28.0	3.02
Texanol	8.0	1.01
Mineral Spirits	30.0	4.70
Deefo 97-2	1.0	.14
Acronal 290 D	76.2	8.76
Water	237.4	28.50
PVC (%)	82.0	
Nonvolatile (%)	52.6	
Nonvolatile by vol. (%)	30.9	
Wt./gal (lb)	12.0	
Visc. (KU)	83±5	

Stipple and Textured Paints

A stippled or textured paint contains silicates such as micas that are suspended in the paint medium. When this type of paint is applied, the silicates impart a textured or stippled appearance to the surface being covered.

Latex Stipple and Texture Compound

	lb	gal
1. Water	416.8	50.00
2. **Daxad 30 (25%)**	11.2	1.17
3. **Everflex GT (55%)**	180.0	20.00
4. Asbestos 7K06	22.5	.96

	lb	gal
5. Mica 325 Mesh	50.0	2.13
6. **Methocel** (65 HG, 4000 DG)	6.0	.53
7. **Duramite**	414.0	18.34
8. **Ti Pure R-901**	50.0	1.49
9. **Celite 281**	100.0	5.22
10. **PMA-18**	3.0	.27
PVC (%)		73.5
Total Solids (%)		59.2
Total Volume Solids (%)		38.8

Procedure:

Charge approximately half of formula water into a change can mixer. Begin agitation and add items 2–5. Dry mix item 6 (**Methocel**) with a portion of item 7 (**Duramite**) and add this with the remainder of item 7 to the mix. Continue agitation until smooth and homogeneous. Add item 8 and mix until smooth again. Slowly add remainder of formula water followed by items 9 and 10. Continue mixing until a homogeneous mix is obtained.

This coating, when applied to dry wall plaster, cement or block walls, will provide a very attractive stipple or textured appearance. It should be applied by brush in heavy coats at full weight (a satisfactory spreading rate is approximately 150 ft^2/gal). The coating, while still wet, should be hand finished with a wad of paper, brush, or damp sponge in a uniform manner to obtain the desired stipple or texture effect.

Alkyd Stipple Enamel

(Semigloss)

	lb	gal
Titanium Dioxide	200	5.89
Gamaco	200	8.90
Zinc **Kadox 515**	75	1.61
Bentone 38	10	.67
Ethyl Alcohol	4	.67
Varkyd 505-70	563	70.40
VM & P Naphtha	130	21.00
Socal 25	80	11.00
Cobalt (6%)	2.5	.31

	lb	gal
Lead (24%)	4.0	.40
Calcium (4%)	2.0	.25
Cosan BBL	7.6	1.00
Blown Linseed Oil	12.0	1.50

Stipple — 15 min set:	Good
N.V. Total (%)	67
PVC (%)	30
Visc.	Heavy-Buttery
Yield (gal)	120
Gloss	60–70 (Fresh)

Acrylic Texture Paint

		lb	gal
1.	**TiPure R-900**	100	2.9
2.	**Duramite**	700	31.2
3.	**100 K Mica**	25	1.1
4.	**Daxad-30**	5	0.5
5.	Ethylene Glycol	28	3.0
6.	**Colloid 606**	2	0.3
7.	**Troysan PMA-30**	0.3	—
5.	**Igepal CO-610**	2	0.2
8.	**Carbopol 934**	1.5	0.1
	Ammonium Hydroxide	8	1.0
	Water	151	18.1
9.	**Rhoplex AC-34** (46% N.V.)	369	41.6
10.	Flint Quartz #2	250	

PVC (%)	66.4
Total Solid (gal) (%)	52.8

Procedure:

To a suitable mixing chamber equipped with an efficient mixer add each of the ingredients in the order given below:

Prepare a pigment slurry by adding all of the water to the mixing chamber followed by **Daxad-30** and ethylene glycol to which **TiPure R-900, Duramite** and **100 K Mica** are added as rapidly as is consistent with efficient mixing. At this point add *just* enough of the **Rhoplex AC-34** binder to give the batch the proper consistency for efficient mixing. Continue to mix until a uniform dispersion is formed, then add the **Colloid 606** (defoamer) and **Troysan PMA-30** (fungicide). While still under agitation (i.e. at a reduced speed) add **Carbopol 934** *slowly* and *carefully* to the mix, alternating it with the addition of **Rhoplex AC-34** in order to prevent the formation of lumps or the solidification of the mixture. When all of the **Rhoplex AC-34** binder has been added, the batch is mised until homogenous. Add the **Igepal CO-610** (surfactant) and ammonium hydroxide. Continue agitation until the ammonium hydroxide has its maximum thickening effect upon the **Carbopol**. Finally, add the flint quartz granules to the finished paint and continue to mix until the quartz is evenly distributed throughout the mass. *Special Note*: The flint quartz granules can be made available for addition on the job and stirred into the paint just prior to application.

Acrylic Resin Emulsion Stipple Paint

		lb	gal
	Rutile Titanium Dioxide	100	2.9
1.	**Duramite**	700	31.1
2.	**100 K-Mica**	25	1.1
3.	**Daxad-30**	5	0.5
4.	Ethylene Glycol	28	3.0
5.	**Colloid 606**	2	0.3
6.	**Troysan PMA-30**	.3	—
4.	**Igepal CO-610**	2	0.2
7.	**Carbopol 934**	2	0.2
	Ammonium Hydroxide	8	1.0
	Water	305	36.7
8.	**Rhoplex AC-55**	207	23.0

PVC (%) 75
Total Solid (gal) (%) 47

Procedure:

To a suitable vessel equipped with an efficient mixer add each of these ingredients in the following order: 1) water; 2) **Daxad-30**; 3) ethylene glycol; 4) rutile titanium dioxide; 5) **Duramite**; 6) **100 K-Mica**.

Mix for an appropriate length of time until a uniformly dispersed slurry is formed. Then add: 7) **Colloid 606**; 8) **Troysan** phenyl mercuric acetate; 9) **Carbopol 934** is added slowly and carefully, because the **Rhoplex AC-55** binder must be titrated with **Carbopol** into the system to prevent solidification of the mixture. About half of the **Carbopol** may be added, before this, binder must be used. After all of the **Rhoplex AC-55** binder is added and the total batch is mixed until homogeneous, add: 10) **Igepal CO-610**; 11) ammonium hydroxide. Continue agitation until the ammonium hydroxide has its maximum thickening effect upon the **Carbopol**.

Chapter IX

VARNISHES, LACQUERS, AND FLOOR FINISHES

Varnishes and lacquers satisfy the demand for a clear finish that protects, and in the case of wood, displays the beauty of the substrate. Exterior wood varnishes easily prevent weathering, but must also prevent ultraviolet light degradation of the wood surface because the resultant deterioration causes a loss of film adhesion. Generally dark-colored (UV absorbent) products are more successful.

Cold-Blend Spar Varnish – Metal

FORMULA No. 1

(Oil)

	lb
CKS-2001 (50% N.V.)	200
Tung Oil	102
Bodied Linseed Oil	160
Mineral Spirits	293

Procedure:

Mix **CKS-2001** resin solution with the oils until uniform. Then thin with mineral spirits.

No. 2

(Oil)

	lb
CKS-2001 (50% N.V.)	200
Copolymer 186	202
Mineral Spirits	137

Procedure:

Mix **CKS-2001** resin solution with the **Copolymer 186** until uniform. Thin with mineral spirits.

No. 3

(Oil)

	lb
CKS-2001 (50% N.V.)	200
Copolymer 186	150
Tung Oil	50
Mineral Spirits	137

Procedure:

Mix **CKS-2001** resin solution with the oils until uniform. Thin with mineral spirits.

Viscosity, Gardner	F-G
Nonvolatile (%)	55.6
Specific Gravity	0.892 (7.44 lb./gal.)
Color, Gardner	4

Performance:

(With the addition of 0.3% lead, 0.03% cobalt, and 0.015% manganese calculated as metal on the weight of oil present; thinned to D-E viscosity with mineral spirits.)

Drying: Set-to-touch (tacky) 1 hr
Slightly tacky 2 hr
Very slight tack 4 hr
Dry 6 hr

Chemical Properties — 72 hr air dry

Resistance to 5% NaOH at 73 F: Passed 24 hr
                                Failed 32 hr

Water immersion at 73 F on tin plate: Slight blush — 24 hr

No. 4

(Oil)

	lb
CKS-2001 (50% N.V.)[1]	200
Tung Oil	147
Bodied Linseed Oil	50
Mineral Spirits	137

Procedure:

Mix **CKS-2001** resin solution with the oils until uniform. Thin with mineral spirits.

---

[1] **CKS 2001** — is a 50% solution of **CKM 2400** in a mineral spirit/isopropanol blend.
CKM 2400 — is a 100% oil-soluble, non-heat–reactive phenolic resin.

No. 5

(Oil)

	lb	gal
CKS-2001 (50% N.V.)[1]	200	26.0
Copolymer 186	150	18.6
Tung Oil	50	6.4
Mineral Spirits	137	21.1

Drying: Set to touch (tacky) 1 hr
        Slightly tacky 2 hr
        Very slight tack 4 hr
        Dry 6 hr

---

[1] **CKS 2001** — is a 50% solution of **CKM 2400** in a mineral spirit/isopropanol blend.
CKM 2400 — is a 100% oil-soluble, non-heat reactive phenolic resin.

General-Purpose Utility Varnish

FORMULA NO. 1

(Oil)

	lb
Tall Oil	1000
Dehydrated Castor Oil (Z-3)	405
Pentaerythritol	136
Maleic Anhydride	40
Mineral Spirits	1530
Zinc Naphthenate	30
Cobalt Naphthenate	10
Manganese Naphthenate	5

NO. 2

(Oil)

	lb
Piccodiene 2215	90
Resinous **Polyol 450** x (1)	10
Oiticica Oil	193
Dehydrated Castor Oil (G-H Visc.)	46
Mineral Spirits	339
Cobalt Naphthenate (6%)	1.5
Lead Naphthenate (24%)	2.0
Zirconium Octoate (6%)	1.5

Cook Procedure:

Charge all oil and resin to kettle and heat to 565 F in 1 hr; cut fire and add Resinous **Polyol 450X**. Hold at 500 F for 20 min then cool and thin. This formulation is oil and gasoline resistant.

Nonvol Solids (%)	50
Color (Gardner)	12
Visc.	G
Dry Time	8 hr
Dry Hard	overnight
Solvent and Oil Resistance after 1 week	
Oil 48 hr	unaffected

Mineral Spirits 18 hr	unaffected
High Flash 2 hr	unaffected
18 hr	softened-recovered

Crown Cap Varnish

(25 gal. oil length)

Piccodiene 2215	100.0
Tung Oil	117.0
Alkali Refined Linseed Oil	39.0
Z-4 Kettle Bodied Linseed Oil	40.0
Mineral Spirits	360.0
Cobalt Naphthenate (6%)	1.6
Iron Naphthenate (6%)	1.6

Cook Procedure:

Heat all resin, tung oil, and alkali refined linseed oil to 540 F, hold for 20 min and check with the Z-4 bodied oil. Thin with mineral spirits at 400 F and add driers.

Wt./gal (lb)	7.25
Visc. (Gardner)	A-2
Recommended Bake	10 min. @ 300 F
Impact Test	Pass 120 in-lb
VM and P Naphtha Immersion 2 hr	Unaffected

Furniture Rubbing Varnish

(Oil)

Congo	130 (100 after running)
Tung Oil	62
Mineral Spirits	195
Zinc Naphthenate	1.0
Manganese Naphthenate	0.2
Cobalt Naphthenate	0.2

Four-Hour Floor and Trim Varnish

(Oil)

Piccolyte S-115 (terpene)	100
Castung 403 (Z-3) (dehydrated castor)	80

Tung Oil	40
Mineral Spirits	220
Cobalt Naphthenate	2
Zinc Naphthenate	5

Clear Floor Varnish

(Epoxy – Ester)

	lb/100 gal
Epotuf 38-406	434.0
Mineral Spirits	216.0
Solvesso 150	74.0
Cobalt Naphthenate (6%)	2.2
Calcium Naphthenate (4%)	1.3
Orthophen 85	0.5

Uses:

Clear finish for wood, concrete and other flooring, clear masonry sealer.

Flatted Alkyd Varnish

FORMULA No. 1

Medium Soya Modified Alkyd	40.0
Mineral Spirits	40.0
Zeothix 60	19.1
Thixcin R	0.9

Procedure:

Grind to Hegman fineness of 5.5–6.0.

No. 2

(Air-Dried)

	Santocel 62 lb	A Competitive Silica lb
Mill Base (quart Mill):		
Dyal XAC-1 (50% Solids) (Soyal Linseed Medium Oil Modified Alkyd)	225	225

	Santocel 62 lb	A Competitive Silica lb
Mineral Spirits	125.45	100
Flatting Agent	28	28
Thixcin R	3.36	—

Procedure:

Pebble Mill: 4 hr, let down as follows and rotate additional 20 min.

Dyal XAC-1 (50% Solids)	221.38	228.10
Mineral Spirits	50	50
Lead Drier (24%)	4.67	4.71
Cobalt Drier (6%)	1.87	1.88
ASA	1.12	1.13
Solids (%)	38.52	39.85
Flatting Agent In Solids (%)	5	5
Hegman Gauge	11.0	11.0
Settling after 3 months	None	(Hard settling gel)

Alkyd Rubber Lacquer

FORMULA No. 1

Titanium Dioxide	30.0
½″ RS Nitrocellulose	24.5
Paraplex RG-2	45.5

No. 2

Titanium Dioxide	23.5
½″ RS Nitrocellulose	34.5
Paraplex G-20	42.0

Epoxy Clear Coating

	lb
Tall Oil **Araldite 7098** Ester (50% N.V.)	724
Cymel 300 Sol'n. (50% N.V.)	80
Catalyst 1010 Sol'n.	4
Xylene	128
n-Butanol	64

	lb
Physical Properties:	
Nonvolatile Content (%)	40
Visc., Stormer (KU)	72
Epoxy Ester/Melamine Formaldehyde Ratio	90/10
Color (Gardner 1933)	3
lb/gal	7.88

Note: Modification of **Araldite** 7098 esters with metallic driers and/or urea-formaldehyde resin can also be used, depending on the required film properties and performance requirements.

Epoxy-Amine Clear Coating

		lb/100 gal
A	**Epotuf 38-501** (6501-75)	508.0
	Xylol	126.0
	Methyl Isobutyl Ketone	105.0
	Cellosolve	23.0
	SR-82	4.0

(Mix 1/2 — 1 hr before use):

B	Triethylene Tetramine	23.0
	Butyl Alcohol	23.0

Uses:

   Air-drying chemically resistant coating for chemical plants, laboratory furniture, etc.

Epoxy Clear Coating

FORMULA No. 1

A	**Araldite 6005**	557.3
B	**Araldite Hardener 830**	178.8
	Araldite Hardener 850	177.9
	Dibutyl Phthalate	38.3

No. 2

A	Araldite 6005	568.1
B	Araldite Hardener 830	298.0
	Araldite Hardener 850	45.1
	Dibutyl Phthalate	40.0

Clear Flexible Epoxy-Amine Exterior Coating

		lb/100 gal
A	Epotuf 37-151	382.0
	No. 840 Silicone Resin	7.0
	Solvesso 150	248.0
	Cellosolve	103.0
B	Epotuf Hardener 37-614	42.0
	Cellosolve	42.0

Uses:

Exterior finish for wood and metal, marine finish, coating for flexible substrates.

Flatting Base

(Polyurethane)

	lb	gal
Arothane 190	458	59.92
Mineral Spirits	229	35.05
MPA-60 (Mineral Spirits)	4	0.57
Flatting Agent*	76	4.46
lb/gal		7.67
Nonvolatile		40%
Pebble mill to 7 grind		

Procedure:

Add the above Flatting Base to the following Clear Varnish at levels of 2–7% flatting agent on vehicle solids to yield varying degrees of sheen.

---

*Santocel FR-C or equal

Clear Varnish

(Polurethane)

	lb	gal
Arothane 190	589	77.10
Mineral Spirits	147	22.52
Manganese Naphthenate (6%)	1.5	0.18
Antioxidant B	1.5	0.20

lb/gal	7.39
Visc.	C
Nonvolatile (%)	40

Gray Floor Enamel

(Polyurethane)

Polurethane 1210	550.0
Titanox RA-45	160.0
Sparmite Barytes	200.0
Lampblack	1.5
Cargill N Lecithin	3.0
Mineral Spirits	122.0
Antioxidant	1.0
Lead Naphthenate (24%)	5.5
Cobalt Naphthenate (6%)	1.1
Manganese Naphthenate (6%)	1.1

MFMA Gym Floor Finish — 40% Solids

(Polyurethane)

Polyurethane 1210	485.00
Heavy Mineral Spirits	125.00
Mineral Spirits	115.00
Antioxidant	1.00
Cobalt Naphthenate (6%)	0.96

General Purpose Varnish — 50% Solids

(Polyurethane)

Polyurethane 1210	625.0

Mineral Spirits	124.0
Antioxidant	1.0
Lead Naphthenate (24%)	5.6
Coblat Naphthenate (6%)	1.3
Manganese Naphthenate (6%)	1.3
Calcium Naphthenate (4%)	3.8

Marine Spar Varnish

(Phenolic)

	lb
Bakelite 9400 (pure phenolic)	100
Castung 103 (G-H) (dehydrated castor)	160
Varsoy (Z-3) (maleinized soybean)	40
Mineral Spirits	247
Zinc Naphthenate	8
Cobalt Naphthenate	3

Chemically Resistant Varnish

(Phenolic)

	lb
Bakelite 254 (pure phenolic)	100
Tung Oil	100
Xylene	198
Zinc Naphthenate	0.2
Cobalt Naphthenate	0.01

Overpaint Varnish

	lb
Krumbhaar 707 (modified phenolic)	100
Tung Oil	40
Refined Linseed Oil	52.5
Cobalt Linoleate	3
Cornstarch	1.5
Short-Range Kerosene	60

Porch and Deck Varnish

	lb
Krumbhaar 505 (modified phenolic)	100
Castung 403 (Z-3) (dehydrated castor)	200
Mineral Spirits	245
Zinc Naphthenate	4.0
Cobalt Naphthenate	1.6

Venetian Blend Varnish

	lb
Krumbhaar 414 (maleic resin)	100
Tung Oil	64
Bodied Linseed Oil (Z)	16
Mineral Spirits	180
Cobalt Naphthenate	0.6

Clear Acrylic Lacquer

A fast-drying clear protective coating.

		% in Aerosol
Concentrate:		55.0
1. Acryloid B72 (40%)	20.0	
2. Toluene	44.0	
3. Methylene Chloride	36.0	
Propellent:		45.0
Isotron 12	100.0	

Procedure:
   Mix ingredients together and fill.

Directions for Use:
   Hold 12–14 in from surface and spray evenly. Invert can and clear valve after spraying.

Package:
   A tinplate container with a paint-type valve, and a gasket suitable for use with methylene chloride.

Precautions:

Warning: Contents under pressure. Do not puncture. Exposure to heat or prolonged exposure to sun may cause bursting. Do not throw into fire or incinerator. Keep from children.

Flammable mixture: Use in a well ventilated area. Do not use near an open flame.

Acryloid Acrylic Wood Lacquer

Acryloid acrylic resins are excellent choices for clear coatings over various substrates. They possess a water-clear transparency and resistance to discoloration even at high temperatures or on exposure to ultraviolet light. They also have outstanding durability and resistance to marring and staining. The following formulation is recommended as a general-purpose lacquer for finishing wood furniture. It has satisfactory print resistance and hardness for this application.

Acryloid B-72	50
½" RS Nitrocellulose	35
Paraplex G-50	15

The formulation has a Gardner-Holdt viscosity of approximately V at 25% soilds. It can be sprayed when reduced with lacquer solvents to about 14% solids.

Protective Acrylic Coating for Copper Metal

The International Copper Research Association (INCRA) has developed an acrylic coating system that can be used to protect copper, bronze or brass surfaces intended for either exterior or interior use. The coating system is based on Acryloid B-44 which provides an excellent balance of hardness, adhesion and film toughness together with film clarity and transparency. It also has excellent resistance to ultraviolet light, and resists yellowing and loss of gloss and clarity upon aging.

Acryloid B-44 (40%)	74.40
Toluene	19.72
Ethyl Alcohol	5.00
Benzotriazole (chelating agent)	0.44
Paraplex G-60 (leveling agent)	0.44
Total solids (%)	30

Procedure:

Reduce to spraying consistency of 11- to 13% solids by slowly adding up to 129.5 parts toluene to the coating as formulated.

Two to three spray coats yield a final dry film thickness of about 1.0 mil. Thirty-minute air-dry periods should be allowed between coats. The finished coating should be air-dried at least 1 hr before handling. Preferably, 48 hr should be allowed for drying after application of the final coat. Brush application is not recommended.

Acrylic Lacquer

Vehicle:

Acryloid B-66	35.0
Xylene or Hydrocarbon Solvent with K.B. Value of approximately 90 and about ½ the evaporation rate of Xylene	65.0

Lacquer:

Roller mill grind or high speed mix

Vehicle	54.0
Santocel 54	0.5
Pigment, D-Series	38.0

Mix with:

Xylene or Hydrocarbon solvent with K.B. Value of approximately 90 and about ½ the evaporation rate of Xylene	7.5

Thin to spraying viscosity with Xylene.

Clear Sanding Sealer

FORMULA No. 1

	lb	gal
Varcopol 474-60 VM	114	15.0
VM&P Naphtha	182	29.0
MPA-60	4	0.5

	lb	gal
Talc **Mistron T-076**	150	6.5
Clay **ASP-400**	175	8.1
Zinc Stearate DLG-20	15	1.7
Amsco Solvent B	21.5	3.0
Varcopol 474-60 VM	230	30.0
VM&P Naphtha	38	6.0
Cobalt Naphthenate (6%)	1.4	0.2
Volatile ASA Exkin #2	0.7	0.1

PVA (%)	39.8
Vehicle Nonvolatile (% Wt.)	35.5
Total NV by Wt. (%)	58.9
Wt. per gal	9.31 #
Visc. (KU)	60–65
Settling	None
Skinning	None
Dry — Set-to-touch	5–10 min
Hard	20–30 min
Sanding after 1½ hr	Excellent
Enamel Holdout	Excellent

No. 2

M-P-A successfully suspends the talc and clay present in this sanding sealer.

	lb	gal
Varcopol 474-60 VM	114	15.0
VM & P Naphtha	182	29.0
M-P-A 60 (Mineral Spirits)	4	.5
Talc **Mistron T-076**	150	6.5
Clay **ASP-400**	175	8.1
Zinc Stearate **DLG-20**	15	1.7
Amsco Solvent B	21.5	3.0
Varcopol 474-60 VM	230	30.0
VM & P Naphtha	38	6.0
Cobalt Naphthenate (6%)	1.4	.2
Volatile ASA Exkin #2	0.7	.1

Varnish

Bakelite Polystrene Resin SKD-3955	20
Toluene	40
Aliphatic Diluent ("Amsco")	40

Maroon Floor Varnish

(Phenolic Resin)

	lb
Synthetic Red Iron Oxide	434.7
Zinc Oxide	23.0
Aluminum Stearate	2.4
CKS-2001* (50% N.V.)	321.0
Raw Tung Oil	150.9
Bodied Linseed Oil (Z-7)	64.2
Mineral Spirits	122.0
Lead Drier (24%)	2.7
Cobalt Drier (6%)	1.1

Procedure:

Grind the pigments in the tung oil, bodied linseed oil, and 42 lb of mineral spirits. Blend into this paste 321 lb of **CKS-2001**, resin solution, and the remaining 80 lb of mineral spirits. Then work in the driers.

---

*Phenolic Resin Sol'n. **CKS-2001** is a 50% Bakelite solution of **CKM-2400** in a mineral spirits/isopropanol blend (85/15). It offers all the advantages of **CKM-2400** in preparing cold-blended varnishes and fortifying alkyds.

Cold-blended varnishes based on **CKS-2001** feature fast-drying speed, durability, and gasproofness. The use with Baker Castor Oil's **Copolymer 186**, which was developed specifically for cold-blended varnishes, permits the formulation of varnishes with outstanding exterior durability.

Shellac Primer Sealer

Titanox RCHT	2,000
Celite 110	250
Shellac 5 # Cut	2,340
Alcohol	400

Furniture Lacquer

½" RS Nitrocellulose	33.4
Amberol 801 Extra Light	33.3
Paraplex 5-B	33.3

Cold-Blend Floor Varnish

(Oil)

	lb
CKS-2001 (50% N.V.)	200
Tung Oil	94
Bodied Linseed	40
Mineral Spirits	76

Procedure:

Mix **CKS-2001** resin solution with the oils until uniform. Thin with mineral spirits.

Oil Length: 17 gal	
Oils: Tung Oil	70%
Kettle Bodied Linseed (Acid No. 8 Visc. Z-7)	30%

Maroon Floor Paint

(Linseed Oil)

	lb
Synthetic Red Iron Oxide	434.7
Zinc Oxide	23.0
Aluminum Stearate	2.4
CKS-2001 (50% N.V.)	321.0
Raw Tung Oil	150.9
Bodied Linseed Oil (Z-7)	64.2
Mineral Spirits	122.0
Lead Drier (24%)	2.7
Cobalt Drier (6%)	1.1

Vehicle:

PC-3468 (17 gal, **CKS-2001**, 70% Tung Oil, 30% Z-7 Bodied Linseed Oil). Cold Blend Varnish

316316316316

ATME

Procedure:

Grind the pigments in the tung oil, bodied linseed oil, and 42 lb of mineral spirits. Blend into this paste 321 lb of **CKS-2001**, resin solution, and the remaining 80 lb of mineral spirits. Then work in the driers.

Note: **PC-3468** is the number assigned to the blend of all the vehicle ingredients. If desired, this can be prepared in advance and used in making the grind.

Gray Floor Enamel

(Epoxy)

A tough, chemical resistant floor paint with excellent adhesion to concrete, wood, brick, metal, and other substrates.

Rutile Titanium Dioxide	250.0
Lampblack	0.5
Antisettling Agent	3.6
SRL-421 Epoxy Ester (60% N.V.)	529.0
Mineral Spirits	153.0
Cobalt Naphthenate (6%)	2.8
Zirconium Octoate (6%)	5.5
Antiskinning Agent	0.9

Dull Finish Varnish

Wood Oil Ester Varnish (50 gal to 100 lb Gum)	20 gal
Steam Distilled Turpentine	10 gal
Magnesium Carbonate	50 lb
Nopco Zinc Stearate H	15 lb

Give light grind on stone or roller mill, and add

Varnish (Type to be determined according to drying, elasticity of finished product)	65 gal
Japan Drier	3–5 gal
Turpentine	3–5 gal

Yields about 105 gal

Dull Finish Without Grinding:

Soft East India Gum	150 lb
Turpentine	25 gal
Mineral Spirits	25 gal
Nopco Zinc Stearate H	5–10 lb
Long Oil Wood-Oil-Rosin Varnish	25 gal

Procedure:

Heat the gum and thinners in a steam-jacketed kettle to 250 F. While hot, strain to remove dirt and bark, being careful, however, not to remove the waxy flattening portion of the gum. To the warm solution, add the zinc stearate which may have been mixed to a thick creamy consistency with a portion of the turpentine. While still warm add the varnish which may be varried to suit the requirements of the finished product. The finished material should be strained through a fine wire screen, or allowed to settle. If the latter method is used the faucet of the tank should be near but not at the bottom of the tank. The entire contents should be drawn from the tank, and thoroughly mixed so that all portions will contain the same amount of flatting agent. This product may be used as a flat varnish or forms the basis for rubbed finishes or eggshell enamels.

In using zinc stearate for producing flat varnishes from varnishes made with synthetic resins, it should not be ground in the varnish but mixed with a portion of the thinners and added to the varnish while it is still warm. If conditions will allow, a small percentage of coal tar thinner, such as toluol or xylol, may be used with zinc stearate to good advantage.

Flooring Primer

(Epoxy)

A	Resypox 1307	70 pbw
B	Resymide 1140	30 pbw
	DMP-30	5 pbw

Seven day cure at room temperature:

Green Floor Enamel

(Epoxy)

	lb/100 gal
RANC Titanox	70.0
Chromium Oxide Green G-6099	140.0
38-406 **Epotuf** (6404-60)	40.0
Mineral Spirits	30.0

Steel Ball — 24 hr

Epotuf 38-406	483.0
Mineral Spirits	145.0
Cobalt Naphthenate (6%)	2.1
Lead Naphthenate (24%)	5.1
ASA	0.9
Paint Additive No. 1	0.9

Uses:

Trade sales and maintenance floor enamel, masonry coatings.

Red Floor Enamel

(Epoxy – Polyamide)

		lb/100 gal
A	**Mapico No. 516** Red Iron Oxide	150.0
	Bentone 27 ⎫ Prewet	4.0
	Ethyl Alcohol ⎭	2.0
	Polyamide Hardener 37-618	216.0
	Solvesso 150	106.0
	Cellosolve	28.0
B	**Epotuf 38-505**	376.0
	Cellosolve	70.0

Uses:

High quality floor enamel for masonry, wood and metal. Suitable for chemical plants, gasoline stations, dairies, breweries, etc.

Chapter X

LUMINESCENT PAINT

Luminescence

Sensations of color produced by conventional pigments or dyes result from the selective absorption of wavelengths in the visible portion of the spectrum. The wavelengths that are not absorbed are reflected or transmitted, giving the sensation of color. This sensation of color is absent when a visible light source is not present, because there is no light energy to reflect or to transmit.

Unlike conventional colorants, luminescent pigments are not primarily light reflectors, but sources of light. They possess the property of absorbing certain types of radiant energy (usually below 4000A and not visible to the eye), converting into longer wavelengths in the visible spectrum, and emitting it as light. The selective wavelengths thus emitted produce the sensation of color or light.

When the emission of luminescent light ceases with the removal of the exciting energy, the luminescent material is called *Fluorescent*. If the emission of light continues for an appreciable period of time after the exciting energy is removed, the luminescent material is known as *Phosphorescent*, and has the property of glowing in complete darkness.

Luminescent Lacquer

(Alkyd)

Pigment (%) 50–60

Luminescent Pigment   100 lb

Vehicle (%) 50–40

1/2 s R.S. Nitrocellulose (dry)	8.25 lb
Alcohol	1/2 gal
Alkyd Resin Sol'n.	1 gal
Dibutyl Phthalate	1/2 gal
Thinner[1]	6 gal

---

[1] Thinner:

Xylol	60%
Ethyl Acetate	17%
Butyl Acetate	15%
Butyl Alcohol	5%
Amyl Acetate	3%

Phosphorescent Ready-Mixed Paint – Short Afterglow Type

(Alkyd)

Pigment (61%)

Phosphorescent – 2301 or 2304	490	lb
Whiting[1]	196	lb
Aluminum Stearate	68	lb

Vehicle (39%)

Alkyd Resin Sol'n.	34.75	gal
Mineral Spirits	34	gal
Cobalt Drier (6% Cobalt)	8	fl. oz.
Calcium Drier (5% Calcium)	16	fl. oz.
Antiskinning Agent	17	fl. oz.

Yield (gal)	100
Wt./gal (lb)	12.50
Nonvolatile in Vehicle (%)	38.2
P.F.V. (%)	58.7

Procedure:

Grind the whiting, aluminum stearate and enough alkyd solution through suitable paint mill. Stir in (do not grind) the phosphorescent pigment with more alkyd and thinner to make a workable paste and add the remainder of the vehicle.

The amount of pigment used in the above formula can be adjusted to meet consistency, application and phosphorescent brightness requirements.

_____

[1] Any standard paint grade of whiting (calcium carbonate) can be used.

Fluorescent Ready-Mixed Flat Paint

(Alkyd)

Pigment (51.5%):

Helecon Fluorescent Pigment	371 lb
Calcium Carbonate [1]	198 lb
Aluminum Stearate	30 lb

Vehicle (48.5%):

Alkyd Resin Sol'n. [2]	43 gal
Thinner	33 gal
Cobalt Drier (6% Co)	18 fl. oz.
Antiskinning Agent	20 fl. oz.

_____

[1] Suspenso or similar whiting.
[2] This formula was based on Glyptal No. 2466.

Baked Phosphorescent Paint for Spray Application

Pigment (21.8%):

Phosphorescent — 2479 [1]	199 lb
Zinc Palmitate	11 lb

Vehicle (78.2%):

Urea–Formaldehyde Resin Sol'n. [2]	91 gal

Procedure:

Add the zinc palmitate to the resin solution and mix thoroughly before stirring in the phosphorescent pigment. The phosphorescent pigmented paint should not be ground in a mill. Only enough zinc palmitate should be used to prevent settling of the pigment during spraying.

Yield (gal)   100

Wt./gal (lb)	9.7
Volatile in Vehicle	50
PFV (%)	16

This paint should be reduced for spraying with a solvent consisting of 80% Xylol — 20% butyl alcohol by weight.

The film should be baked for 1 hr at 250 F.

---

[1] This paint has been designated for commercial application. The amount of pigment used in this paint, whether Phosphorescent-2479, as shown, or other of U.S. Radium's phosphorescent pigments, should be adjusted to meet application requirements.

[2] Beetle 592-8 (resin 50%, butonol 25%, xylol 25%).

Pigments

There are several basic pigment series available from the **Day-Glo Color Corp.** that can be used to formulate aerosol paint systems. These pigments are A Series, D Series, AX Series and T Series.

For all practical purposes, the **A** and **D Pigments** can be treated as one series, as the basic difference is that the **D Series** contains an ultraviolet absorber to increase the outdoor lightfastness. The color difference between **A** and **D Pigments** is minimal in most cases. In formulating fluorescent aerosol paints, lightfastness is generally taken into consideration and the D Series is normally used; although from a cost standpoint, the **A Series** would produce a paint that would normally be satisfactory and slightly cheaper.

Fluorescent Concentrate Formula Using A or D Series Pigments

(Acrylic)

	lb/100 gal
Acryloid F-10	373.1
(40% Solids in 9:1 Mineral Thinner/**Amsco F**)	
A or D Series Day-Glo Pigment	287.0
Lactol Spirits	145.1
Toluene	14.8

Procedure:

Disperse the pigment in the vehicle using a high speed disperser, such as a Cowles Dissolver. After the pigment is thoroughly wet down and dispersed, add the solvent under slow speed agitation. The concentrate should be sieved or strained to ensure that no agglomerates or foreign particles are present. The resulting product should have a flat finish with a typical viscosity range of from 800 to 1400 Centipoise, a Hegman scratch of 7 or better and a weight per gallon of about 8.0–8.3 lb/gal.

One hundred parts of this concentrate can be reduced, after it has cooled to room temperature, to canning viscosity by adding 100 pbw of toluene or a mixture of one-third toulene and two-thirds Lactol Spirits. This reduction will give a viscosity of about 10–14 s using a number 4 Ford cup. The reduced material is now ready to can with a propellant such as Freon 12 using 55% of the thinned product and 45% of the propellant by weight.

An alternative canning procedure with a lower propellant cost that can be considered is:

Concentrate	30.1
Lactol Spirits	10.4
Toluene	5.3
Isopropane	27.1
Freon 11	27.1

Formula Using Day-Glo AX Series Pigment

The following formula has a lower raw material cost than the preceding formulas due to the use of inexpensive colorless extending pigments in conjunction with "Extra Strength" **AX Pigment**. The color effects achieved are similar but there is some sacrifice in resistance to direct sunlight.

(Acrylic)

Acryloid B-67 (45% solids in VM&P)	30.0
AX Series Day-Glo Pigment	17.3
Atomite	34.3
Cab-O-Sil H-5	0.4
Aluminum Stearate Gel	3.0
(20% Plymouth Parsons 222-2B in VM&P)	
VM&P Naphtha	15.0

Procedure:

Using a high speed disperser such as a Cowles, disperse in order listed. Be sure that a strong vortex is maintained throughout the additions.

This formula can be reduced with aliphatic solvents before canning. The use of **Buna** gaskets is recommended over **Neoprene** gaskets, as they have proven to be more reliable. The lower cost aliphatic propellants should be considered for use with this formula, as well as the fluorocarbons.

Methylene Chloride Formula Using Day-Glo T Series

It should be noted that chlorinated solvents such as methylene chloride should **never** be used in any of the preceding formulas based on thermoplastic **A, D** or **AX Series** Pigments. Recent experiments in the Day-Glo Laboratories, however, indicate that it is possible to formulate methylene chloride compatible aerosol paints by using the more solvent resistant, thermoset **T Series Pigments**. A starting formulation for your consideration follows. It is a concentrate that can be thinned with methylene chloride and canned with a hydrocarbon propellant.

Acryloid F-10	44.6
Post-4	1.2
Day-Glo T Series Pigment	34.8
Lactol Spirits	17.6
Toluene	1.8

Procedure:

This concentrate may be prepared on a high-speed dissolver such as a Cowles. Materials should be added in order listed and each one dispersed thoroughly before making next addition.

It is strongly suggested that this formula or modifications of it be thoroughly checked for incubation and can stability before being put into production.

Vehicles

It has generally been found that acrylic vehicles are best for fluorescent aerosol systems. They wet the fluorescent pigments adequately and produce a relatively hard finish when dry. They lend themselves to fast evaporating solvents which are necessary for aerosol paint systems. Other vehicles that can be considered are fast drying vinyl-toluene alkyds in suitable aliphatic solvents such as **Aropol 832-V-50**.

Two acrylic vehicles which have proven to be quite successful are **Acryloid F-10** and **Acryloid B-67**. A third acrylic vehicle may be produced by dissolving 40.0% **Elvacite 2044** in an aliphatic solvent, such as Lactol Spirits.

Solvents

Aliphatic hydrocarbons have been found to cause the least amount of problems in fluorescent aerosol systems, as they normally have no effect on the fluorescent pigments. Aromatic hydrocarbons, such as xylene or toluene may be used if the temperature is kept below 90 F during processing of the concentrates. High concentrations of aromatic solvents at elevated temperatures will cause the fluorescent pigment to agglomerate and cause serious problems in the spray nozzle.

Ketones, esters, alcohols and highly active solvents are not recommended, as they normally will partially or completely dissolve **A**, **D** and **AX Series Pigments**.

The use of **Buna** gaskets is recommended over **Neoprene**, as they have proven to be more reliable with the solvents and resins normally used.

Propellants

The "classic" propellant for fluorescent aerosol systems has been **Freon 12**. This propellant has proven to be quite satisfactory for use with fluorescent aerosol paints. In recent years it has been found that other propellants, such as propane and isobutane can be used providing the vehicle system used will tolerate reduction with the aliphatic propellants. The system must be "rich" enough to allow for the use of the relatively low KB propellants.

Methylene chloride and Freon 11 have also been used, but *only* where the pigment is the **T Series**, as the **A Series** and **D Series** are completely soluble in the methylene chloride. A thorough test should be made if methylene chloride and **Freon 11** are used.

Chapter XI

SPECIALTY PAINTS

These formulas are representative of an expanding demand for coatings that possess unique characteristics.

Book Cloth Coating

Roller mill grind:

Chrome Green	33.3
Paraplex 5-B	22.2

Mix with:

5″ to 6″ RS Nitrocellulose	26.6
Paraplex 5-B	17.9

Ratio of **Paraplex 5-B** to nitrocellulose 1.5/1. Reduce with the following thinner:

Ethyl Acetate	33%
Ethyl Alcohol	30%
Troluoil	32$
Acetone	5%

Upholstery Fabric Coating

Roller mill grind:

Pigment	16.7
Paraplex RG-8	8.3

Mix with:

15″ to 20″ RS Nitrocellulose	33.3
Paraplex RG-8	41.7

Ratio of **Paraplex RG-8** to nitrocellulose 1.5/1. Reduce to 50% solids with the following thinner:

Ethyl Alcohol	50%
Ethyl Acetate	33%
Butyl Alcohol	17%

Green Epoxy-Polyamide Flexible Fire-Retardant Coating

	lb/100 gal
A **Titanox RA**	57.0
Antimony Oxide	185.0
Hansa Yellow	3.0
Monastral Green GI-674D	0.3
Imperial IAF Compound	2.0
Cab-O-Sil	19.0
Nytal 300	244.0
Dri-Film 103 (10% in Xylol)	1.7
No. 840 Silicone Resin	7.0
Epotuf 37-200	151.0
Epotuf 37-151	151.0
Arochlor 1248	103.0
Solvesso 150	180.0
Pine Oil	26.0

Pebble mill to 5 N.S. fineness

B Polyamide Hardener 37-620	85.0
n-**Butanol**	28.0

Uses:
Fire-retardant mastic for foams and other flexible substrates.

Intumescent Fire Retardant Coating

	Ball Mill	Cowles
		lb/100 gal

Add:

Mineral Spirits[1]	326	225
Chlorinated Paraffin 70	80	80
Pliolite VTAC-L	100	100

Add and grind:

Titanium Dioxide RANC	90	90
Dipentaerythritol	75	75
Melamine	75	75
Phos-Check P/30	250	250
Bentone 38	5	5

Stir in:

Mineral Spirits[1]	—	101
Methanol	2	2

PVC (%)	65.1
Wt. Solids (%)	67.2
Vol. Solids (%)	50.0
Visc. (KU)	93

For brush or roller application, no reduction is necessary. For air spray, reduce one gallon of material with one pint of VM&P Naphtha[2]. Airless spray requires no reduction.

---

[1] In areas subject to air pollution regulations, either **Amsco Mineral Spirits 20-H, Chevron Thinner 350 H**, or equivalent solvents (minimum 36 KB) may be substituted for mineral spirits in the formulations.
[2] **Ashland VM&P Naphtha-K, Chevron Thinner 298, Napoleum 50-S**, or equivalent may be substituted for VM&P Naphtha to meet air pollution regulations.

Fire Retardant Clear Topcoats

Formula	No. 1	No. 2	No. 3
	(Flat)	*(Semigloss)*	*(Gloss)*
		lb/100 gal	
Mineral Spirits[1]	165	162	163
VM&P Naphtha[2]	380	375	377
Pliolite VTAC-L	76	81	85
Chlorinated Paraffin 70	76	81	85
Firemaster T-23-P	8	8	10
Celite 499	20	10	—
Solids (%)	25.0	25.0	25.0

Clear topcoats may be applied by brush, air spray, roller or airless spray.

---

[1] In areas subject to air pollution regulations, either **Amsco Mineral Spirits 20-H, Chevron Thinner 350 H**, or equivalent solvents (minimum 36 KB) may be substituted for mineral spirits in the formulations.

[2] **Ashland VM&P Naphtha-K, Chevron Thinner 298, Napoleum 50-S**, or equivalent may be substituted for VM&P Naphtha to meet air pollution regulations.

"Hypalon" Paint for Flexible Substrates

(Latex)

Base Lacquer:

Hypalon 20	100
Staybelite Resin	2.5
Cellosolve Solvent	20
Xylene	205
Ionol	2

Dispersion:

Tri-Mal	40
Ti-Pure R-900	80
Atomite Whiting	30
MBTS	1

Nuosperse 657	2
Cellosolve Solvent	30
Xylene	112
MPA 60 (Xylene)	20

Wt./gal	9.29 lb
Coverage/gal/mil	360 ft^2
Solids by Wt. (%)	40
Solids by vol (%)	23-4
Consistency (cps)	3500
PVC (%)	31.8

"Hypalon" Paint for Rigid or Semirigid Substrates

(Latex)

		lb/100 gal
Solution:		
Hypalon 30	50	82.5
LD-255	50	82.5
Staybelite Resin	1.5	2.5
Ionol	2	3.3
SF-69 (2% in xylene)	2	3.3
Xylene	25	41.3
Toluene	145	239.5
Cellosolve Solvent	15	24.8
Dispersion*:		
Tri-Mal	40	66.1
Thiuram M	0.3	0.5
MBTS	0.5	0.8
Ti-Pure R-900	70	115.5
Atomite Whiting	25	41.3
MPA 60 Xylene	15	24.8
Nuosperse 657	1	1.7
Xylene	120	198.0
Cellosolve Solvent	15	24.8
Solids (Wt. %)	45	
Solids (vol. %)	24	
Visc. (cps)	1130	

PVC (%)	32

---

*Grind to Hegman Fineness of 8. After grinding, add the B dispersion to A solution under slow agitation.

Ignition Waterproofing Seal Coating

(Acrylic)

A fast-drying clear coating to keep ignition systems free of moisutre.

		% in Aerosol
Concentrate:		50.0
1. **Acryloid A-101**	5.0%	
2. **Acryloid B-82**	27.0%	
3. **Kesscoflex DBT**	2.0%	
4. Toluene	16.0%	
5. Methylene Chloride	50.0%	
Propellent:		50.0
Isotron 12	80.0%	
Isotron 11	20.0%	

Package:

Tinplate container with a paint-type valve, and a gasket suitable for use with methylene chloride.

Procedure:

Dissolve the resins and **Kesscoflex DBT** in the solvents and fill.

Directions for Use:

Hold can 12–14 in. from surface and spray evenly. Invert can and clear valve after spraying.

Precautions: Warning:

Contents under pressure. Do not puncture. Exposure to heat or prolonged exposure to sun may cause bursting. Do not throw into fire or incinerator. Keep from children.

Combustible mixture: Use in a well ventilated area. Do not use near an open flame.

Polyurethane Paper Coating

FORMULA NO. 1

Polyol Sol'n.:

Desmophen 1300	232
Vinylite VAGH	89
Solvent Blend*	507

Polyisocyanate:

Desmodur IL	107
Mondur CB-75	65

---

*ethyl acetate/butyl acetate/2-ethoxyethl acetate 1:2:1

Notes:

1. Paper Coating PCM-2, because of **Desmodur IL** content, will show faster curing response than film based on **Mondur CB-75** alone.

2. Coating will reach tack-free state within 30 s at 120 C.

No. 2

		Wt.	Solids
A	Polyol Solution		
	Desmophen 1300	233	175
	Vinylite VAGH	90	90
	Solvent Blend*	510	
B	Polyisocyanate		
	Mondur HC	167	100

---

*ethyl acetate/butyl acetate/2-ethoxyethyl acetate 1:2:1

Notes:

1. The gloss of the coating can be increased by a decrease in the portion of VAGH used.

2. Scratch resistance of the coating can be increased by addition of additives, such as silicones.

3. Equivalent weight of VAGH taken as 680 or 2.5% hydroxyl.

4. Because coating is based on **Mondur HC**, film will show overall good film properties along with good nonyellowing characteristics.

5. Coating will be tack-free within 30 s at 120 C.

Fluorescent Gravure Ink

(Acrylic)

Acryloid NAD-10	30.0
Fluorescent Pigment	45.0
Toner	1.0
VM & P Naphtha	13.0
Toluene	1.0
Thickener (100% solids)	1.0
Textile Spirits/Toluene (9/1)	9.0

Alkyd Roof Paint

FORMULA No. 1

(Green)

	lb/100 gal
Chromium Oxide Green	200
Thixcin R	3
Haynie 111-70 Alkyd	482
High Boiling Mineral Spirits	93
Mineral Spirits	116
Cobalt (6%)	4.5
Calcium (5%)	3.5
24%	7
Antiskinning Agent	1

To prevent seeding of **Thixcin R** do not allow processing temperature to exceed 130°F.

lb per gal	9.10
Visc. (KU)	67–75
Grind	5

No. 2

(Red)

	lb/100 gal
Red Oxide (99% Iron)	150
Thixcin R	3.75

	lb/100 gal
Haynie 111-70 Alkyd	531
Mineral Spirits	65
High Boiling Mineral Spirits	103
Cobalt (6%)	5
Calcium (5%)	3.75
Lead (24%)	7.75
Antiskinning Agent	1
lb per gal	8.69
Visc. (KU)	80–90
Grind	5
Approximate RMC	$1.01

Asphalt Roof Paint

(Polyvinyl Chloride)

	lb	gal
Premix:		
Water	125.0	15.00
Daxad 30 Dispersant	10.0	1.17
Tergitol NPX Surfactant	5.0	0.56
Victawet 35-B Surfactant	3.0	0.31
Increase speed and add slowly:		
Chemacoil TA-100	78.0	10.00
Adjust speed to disperse pigments:		
202 Basic Silicate White Lead	100.0	1.87
Ti-Pure R-901 Titanium Dioxide	125.0	3.65
Duramite Calcium Carbonate	100.0	4.46
Mica, 325-mesh, water-ground	25.0	1.05
Ethylene Glycol	18.5	2.00
Nuodex PMA-18 Fungicide	3.0	0.27
Colloid 581-B Defoamer	2.0	0.30
Slow to mixing speed:		
Cellosize QP-15000 (2%) Thickener	41.6	5.00
Everflex BG Vinyl Acetate Copolymer Emulsion (52%)	360.0	40.44

	lb	gal
Super-Cobalt Drier	1.0	0.13
7F Asbestos Fiber	50.0	2.35
Water or Thickener Solution	125.0	15.00

Smooth and pasty; easy to brush.

PVC (%)   44.4

Chemacoil TA-100 = 30% of binder solids

Synthetic Rubber Roof Coating

(Latex)

	lb/100 gal
Mill Base Dispersion: (ball mill, 138°F):	
Tribasic Lead Maleate	58.20
Titanium Dioxide	116.20
Calcium Carbonate	43.55
MBTS	1.56
Zalba	2.91
Dispersion Aid	2.91
2-Ethoxyethanol	43.55
M-P-A 60 (Xylene)	33.0(B)

Let down Base Lacquer:	
Hypalon 20	145.2
Hydrogenated Wood Rosin	3.63
2-Ethoxyethanol	29.1
Xylene	298.13

Splash Zone Coating

FORMULA NO. 1

(Epoxy)

	lb
A **Araldite 6005**	214.1
Araldite 6010	214.1
Titanium Dioxide	64.4
Silica Flour	517.4

	lb
Asbestos Fiber	19.8
Antisag Agent	31.7
B **Araldite Hardener 830**	127.9
Araldite Hardener 850	127.9

No. 2

	lb	gal
Resypox 1628	597	58.1
Mod-Epox	55	5.6
Magnesium Oxide	trace*	trace
Mineral Filler 1719	75	3.5
Silica 219	701	31.8
Medium Chrome Yellow Y-469-D	49	1.0
Resymide 1125	542	66.2
Mineral Filler 1719	542	66.2
Silica 219	574	26.0
Lampblack BTA	3	0.2

Procedure:

Stabilize **Mod-Epox** with magnesium oxide, add **Resypox** to mixer and slowly add pigments. Blend until smooth. Give one loose pass on a three (3) – roll mill. Add **Resymide** to mixer and add pigments slowly. Blend until smooth. Give one loose pass on three (3) – roll mill.

Mixing:

Add 1.3 volumes of **Resymide** to 1.0 vol of **Resypox** and mix to a uniform olive color.

Application:

Sandblast to 1–2 ft below low tide level. (Procedure not troublesome under water.) Use gloves and apply by hand 1/8–1/4 in. thick. Apply in ring above water line and smear up and down over area to be covered. Feather edges.

---

*1% by Wt. based on Mod-Epox.

Stencil Paint

(Acrylic)

A fast-drying, low-gloss coating suitable for stencilling cartons and boxes.

		% in Aerosol
Concentrate:		50.0
Acryloid B72 (40%)	20.0	
Pigment Dispersion	10.0	
Kesscoflex DBT	1.5	
Toluene	8.5	
Methylene Chloride	60.0	
Propellent:		50.0
Isotron 12	100.0	

Package:
Tinplate container with a paint-type valve and a glass marble or steel ball agitator. Gasket suitable for use with methylene chloride.

Procedure:
Dissolve the first three ingredients in the methylene chloride, then add the toluene. Filter before using.

Directions for Use:
Shake well. Hold can 12–14 in. from surface to be sprayed.

Precautions: Warning:
Contents under pressure. Do not puncture. Exposure to heat or prolonged exposure to sun may cause bursting. Do not throw into fire or incinerator. Keep from children.

Nonskid Paint for Decks and Swimming Pool Borders

(Chlorinated Rubber)

Alloprene 10	8.2
Cereclor 42	2.2
Cereclor 70	8.2
Chromium Oxide	15.0

Barytes	27.0
Thixcin R	2.2
Xylene	22.2
Sand — Mesh 80 — 100	15.0

The above nonskid paint tends to be porous because of the sand content so it is usual to apply it on top of a primer paint when steel surfaces have to be protected.

Swimming Pool Paint

FORMULA No. 1

(Nonreducible — Acrylic Latex)

lb/100 gal

Paste:

Titanox RANC	175
ASP 105	75
Phthalocyanine Blue BT 449-D	2
Bentone 11	4
Pliolite ACL	60
Xylene	240

Let down:

Pliolite ACL	147
Dow 276-V2	18
Xylene	230
PVC (%)	25.7
Solids (%)	50.6
KU	65–70
Gloss	70–75
Hiding Power	450 ft^2/gal

No. 2

(Flat – Acrylic Latex)

	(White) lb/100 gal	(Blue) lb/100 gal	(Green) lb/100 gal
Titanium Dioxide-Rutile NC	175	150	150
Barytes	361	380	380
Mica	80	80	80
Diatomaceous Silica	80	80	80
Phthalocyanine Blue	–	2	–
Phthalocyanine Green Toner	–	–	2.5
Thixcin R	5	5	5
Soya Lecithin	3	3	3
Pliolite S-5	89	88	88
Chlorinated Paraffin (40%)	38	38	38
Chlorinated Paraffin (70%)	38	38	38
High-Flash Naphtha	205	205	205
Mineral Spirits	205	205	205
Solids (%)	67.9	67.8	67.8
PVC (%)	57.6	57.2	57.2
Visc. (KU)	73–78	73–78	73–78

No. 3

(Blue – Flat)

	lb/100 gal
Titanox RANC	150
Barytes #1	275
Mineralite 3X	50
Celite 110	65
Phthalocyanine Blue BT449-D	2
Thixcin	5
Soya Lecithin	6
Pliolite ACL	120
Chlorinated Paraffin (40%)	40
Chlorinated Paraffin (70%)	40
Solvesso 100	208
Mineral Spirits	208

	lb/100 gal
PVC (%)	45.6
Solids (%)	64.4
Visc. (KU)	68–72
Gloss	5
Hiding Power	475 ft.^{2}/gal.

Traffic Line Paint

FORMULA No. 1

(Yellow – Oil)

	lb/100 gal
Whiting	178.5
Chrome Yellow (Medium)	230.0
Magnesium Silicate	50.0
Diatomaceous Silica	80.0
Mica	60.0
Propylene Oxide	2.0
Varnish (50% Nonvolatile)[1]	247.0
Chlorinated Paraffin	42.0
Chlorinated Rubber	86.0
Toluene, TT-T-548	173.2
Aliphatic Thinner	76.8
Cobalt Naphthenate (6%)	1.3
Lead Naphthenate (24%)	2.5
Antiskinning Agent	1.2
Antisettling Agents†	7.0 max.

---

[1] The varnish is based on 15 gal oil length bodied dehydrated castor oil and petroleum hydrocarbon resin.

† **M-P-A** (Toluene) at a level of 7 lb PHG is recommended in this specification to insure against caking and excessive settling of the pigment in the package as well as to provide for control of the viscosity. If **M-P-A 60** (Toluene) is used, a level of 12 lb PHG is recommended.

No. 2

(Yellow — Alkyd)

	lb/100 gal
Medium Chrome Yellow	200
Calcium Carbonate	390
Talc	115
MPA	5.5
Haynie 151-T-60 Alkyd	248
Parlon S-10	43
Epichlorohydrin	1.34
Toluol	236
Mineral Spirits	31
Antiskinning Agent	1.34
lb per gal	12.71
Visc. (KU)	68–77
Pigment by Wt. (%)	55.5
PVC (%)	58
Nonvolatile in Vehicle (%)	34.2
Total Nonvolatile (%)	71
Grind	3
Dry	no traffic pickup 15 min

	No. 3	No. 4
	(White — Alkyd)	
Titanox C-50	39.00	39.00
Lesamite Whiting	16.40	16.40
Thixcin R	0.20	0.20
Alkyd Resin (60% Solids)	27.10	27.10
Paralon S-10 (50% in Toluene)	3.60	1.80
Lead Naphthenate (24%)	0.77	0.77
Cobalt Naphthenate (6%)	0.20	.20
Toluene	12.60	12.60
Epichlorohydrin	.08	0.08
Antiskinning Agent	.05	0.05
Velsicol X-37 (50% in Toluene)	—	1.80

	No. 3 (cont'd)	No. 4 (cont'd)
	(White – Alkyd)	
Wt./gal	13.2	13.1
PVC (%)	56.0	56.1
Nonvolatile (%)	73	73
Visc. (KU)	76	72

No. 5

(Latex – Low Cost)

	lb	gal
Rutile Titanium Dioxide	50	1.43
Rutile Titanium Calcium (30%)	300	11.10
Asbestine 3x Talc	75	3.24
Celite 281	75	⁻3.75
Mica (325 mesh water ground)	75	3.25
Kelecin F	4	0.50
Keltrol 1074	375	49.50
6% Cobalt Naphthenate (0.03%)	1	0.12
Toluol	230	31.80

Visc. (KU)	75
PVC (%)	46.2
Vehicle Solids (%)	37.2
Total Solids (%)	67.5
Wt./gal (lb)	11.5

No. 6

(White – Chlorinated Rubber)

Alloprene (20 cps)	3.8
Soya Alkyd Resin[1]	25.3
Titanium Calcium Pigment	46.2
Paris White	5.7
Asbestine	5.7
Thixcin E	0.2
Epichlorohydrin	0.1
Benzene	13.0

Visc. 180 s	2.5 poise
PVC	54.5%
No pick-up time	10 min

Baker Comment:

    Thixcin E can be replaced by **Thixcin R** which is more commonly used in the paint industry.

---

[1] Medium pentaerythritol phthalic type containing minimum of 30% phthalic anhydride and a minimum of 54% soybean oil acids. The alkyd is supplied as a 60% solution in toluene.

No. 7

(Chlorinated Rubber)

Parlon S-20	3.3
Becksol 96	13.1
Medium Chrome Yellow X-2541 Imperial	21.3
Magnesium Silicate, **Asbestine XXX**	7.3
Calcium Carbonate, **Atomite**	32.4
M-P-A additive	0.6
Epichlorohydrin	0.1
Antiskinning agent	0.1
Cobalt Naphthenate (6%)	0.1
Lead Naphthenate (24%)	0.2
Toluene	16.5
Benzene	5.0
Dry-time (min)	17
Solids, as sprayed %	72
Wet Film thickness, as applied	14–16 mils
Visc. (KU)	75
Settling:	
1 month	nil
8 months	sl.

Performance:

Durability (% Intact)

Concrete:

4 months	96
6 months	—
7 months	92
8 months	78

Asphalt:

4 months	97
6 months	—
7 months	92
8 months	82

Baker Comment:

We recommend that the 0.6 parts of **M-P-A** be replaced with 1.0 part of **M-P-A 60** (Toluene) for more effective dispersion.

No. 8

(Yellow – Chlorinated Rubber)

Parlon S-20	3.3
Beckosol 96	13.1
Medium Chrome Yellow	21.3
Magnesium Silicate, **Asbestine XXX**	7.3
Calcium Carbonate, **Atomite**	32.4
M-P-A additive	0.6
Epichlorohydrin	0.1
Antiskinning Agent	0.1
Cobalt Naphthenate (6%)	0.1
Lead Naphthenate (24%)	0.2
Toluene	16.5
Benzene	5.0
Dry-time (min)	17
Solids, as sprayed (%)	72
Wet Film thickness, as applied	14–16 mils
Visc., as applied, KU	75
Settling:[1]	
1 month	nil.
8 months	sl.

Performance:

Durability (% Intact)

Concrete:

4 months	96
6 months	—
7 months	92
8 months	78

Asphalt:

4 months	97
6 months	—
7 months	92
8 months	82

Baker Comment:

We recommend that the 0.6 parts of **M-P-A** be replaced with 1.0 part of **M-P-A 60** (Toluene) for more effective dispersion.

---

[1] Determined by stirring with a spatula and visual observation

Tree Wound Paint

A protective coating for horticultural use.

		% in Aerosol
Concentrate:		70.0
Surett Fluid #50	72.0	
VMP Naphtha	28.0	
Propellent:		30.0
Isotron 12	100.0	

Package:

0.5 lb tinplate container. Valve with .04 spray orifice.

Procedure:

Dissolve the Surett Fluid in the VMP Naphtha.

Directions for Use:

Spray on plant surfaces exposed by storm damages or as a result of pruning to form a moisture and insect-resistant coating.

Precautions: Warning:

Contents under pressure. Do not puncture. Exposure to heat or prolonged exposure to sun may cause bursting. Do not throw into fire or incinerator. Keep from children.

Flammable mixture: Do not use near an open flame.

Acrylic Lacquers

Naval Weapons Coating

This specification covers a grade of acrylic lacquer for use as a general-purpose exterior protective coating for the metal surfaces on naval weapons. It is particularly formulated for resistance to diester lubricating oils and heat, and is primarily for spray application. The lacquer is intended for use over a system of epoxy–polyamide primer **MIL-P-12277**, with or without pretreatment coating **MIL-C-8514**. It may also be used as an insignia and marking lacquer directly over freshly applied epoxy–polyamide topcoat MIL-C-22750.

White Gloss

Rutile Titanium Dioxide	18.38
Acryloid A-21 (30% in toluene/butanol, 90/10)	43.65

Automotive Refinishing

Auto refinishing applications require an air-dry or force-dry system with high gloss and excellent exterior durability. This same type of coating would also be excellent for finishes on farm equipment and lawn mowers, on new truck, trailer and railcars and for maintenance finishes. The following formulations are alkyd-acrylic systems based on **Acryloid B-99** and **Amberlac 292X**. **Acryloid B-66** is also used in this type application. A variety of colors and both ball mill and roller mill grinds are shown.

Acrylic Automotive Refinish

Formula	No. 1	No. 2	No. 3	No. 4
	(Blue)	(Blue Metallic)	(Red)	(Black)
Ball mill grind:				
Excelsior Black	—	—	—	28
Monastral Blue BT-413D	40	40	—	—
Molybdate Orange YE-421D	—	—	90	—
Monastral Violet RT-819D	—	—	10	—
Acryloid B-99 (50%)	120	120	200	120
Xylene	120	240	100	240

Decorative Spray Paint

A decorative coating for use in the preparation of floral displays and special decorative effects. Formulated to be nondestructive to polystyrene foam.

		% in Aerosol
Concentrate:		57.0
Ethyl Cellulose N10	4.5	
Cellolyn 102	4.5	
Dibutylphthalate	1.0	
Ethanol	15.7	
Butanol	10.5	
Cellosolve	5.3	
Butyl Cellosolve	6.3	
Butyl Acetate	6.8	
Toluene	7.9	
Acetone	37.5	
Propellent:		43.0
Isotron 12	100.0%	

Package:

Tinplate container with a paint-type valve.

Procedure:

Dissolve the resinous materials in the solvents with agitation until a clear solution is obtained. Add pigment dispensions if desired. May be cold or pressure filled.

Note:

Pigmentation may be accomplished by substituting a pigment paste ground in ethyl cellulose for 2% of the concentrate. This percentage can be varied depending on the hiding or tinting desired. Satisfactory sprays were obtained when bronzing powders were added in the amount of 2.5%. Below are listed several pigmenting materials

Phthalocyanine Blue Paste
Phthalocyanine Green Paste

Directions for Use:

Spray lightly on surface to be coated. Allow to dry between successive coats.

Precautions: Warning:

Contents under pressure. Do not puncture. Exposure to heat or prolonged exposure to sun may cause bursting. Do not throw into fire or incinerator. Keep from children.

Epoxy Coating for Rigid Substrates

	FORMULA	No. 1	No. 2
		lb/100 gal	
		(White)	(Gray)
A	Epon 815	523	520
	Lampblack	–	2
	Titanox Ranc	122	79
	Sparmite	123	189
	Celite 165-S	48	48
	Silicone Resin SR-82	8	8
	Phenol Sol'n. (50% Wt. in Neosole)	31	31
	Thixatrol ST	9	10
B	Epon Curing Agent E-3	230	209

The Thixatrol ST, titanium dioxide, Sparmite, Celite 165-S should be dispersed in approximately one-half of the **Epon 815** in a disperser to a fineness of at leat 6H (Hegman). The remaining ingredients are then added with good agitation. B is added just prior to use.

Application:

Spray: In this system the curing mechanism involves activation of the curing agent by atmospheric moisture. Normally, the coating must be exposed to the atmosphere for about a day before sufficient moisture is picked up by the coating to activate all of the curing agent. The need for an excellent sag and flow agent is therefore apparent.

Recommended for use over rigid substrates where appearance (as well as protection) is important, particularly tank linings, masonry and marine coatings.

Chapter XII

INDUSTRIAL PAINTS

These are coatings that are used on rubber, wood, paper, plastic, metal, glass, and any other type of manufactured object. Required characteristics are as varied as the market itself — from photosensitive materials to floral patterned wallpaper. They may be decorative or integrally involved in the product's function or both. Most, however, are designed for the large-scale, specific application techniques available in factories.

Aluminum Baking Enamel

(Alkyd)

	lb	gal
MO-588 Aluminum Paste	25.0	1.77
Duramac 2483 Alkyd Resin (50%)	450.0	54.54
Cymel 248-8 Melamine Coating		
Resin (55%)	55.0	6.55
Uformite F-200E Urea Coating		
Resin (50%)	90.0	10.71
Butanol	50.0	7.46
V M & P Naphtha	58.0	9.20
Xylene	62.0	8.58
Adamac Catalyst 20	9.0	1.23
Visc. (#4 Ford cup) — s	23	
Wt. per gal (lb)	8.0	
Total Solids (% by Wt.)	39.65	

Nonvolatile Vehicle	75% alkyd
	10% melamine
	15% urea
Catalyst Level:	3% on NVV
Baking schedule	15 min at 250 F
range:	2 min at 350 F

Properties on **Bonderite** steel panel baked 15 min at 250°F:

Film Thickness	0.9 min
Pencil Hardness	2H
60° Gloss Reading	90
Impact, direct	20 in.–lb
reverse	4 in.–lb
Adhesion	100%
Flexibility on Mandrel	OK

Gloss Black Enamel

(Alkyd)

Ball mill base:

Aroplaz 2462-XM-50 or	
Aroplaz 2480-XB-50	
(50% resin sol'n.)	40.0
Neospectra Mark 2	2.0
n-Butanol	4.0
Xylene	20.0
Methyl Ethyl Ketone	2.0
Isophorone	2.0

Mix-off:

The following mixture is added with stirring to mill base:

G-E 75108 Methylon Resin	20.0
10% H_3PO_4 in Ethyl Acetate	5.0 (1.25% H_3PO_4 on resin solids)
Ethyl Acetate	5.0

Resistant to solvents, chemicals, soap solutions, salt solutions, etc. Excellent flexibility.

Resin Solids (%)	40.0
Visc. (#4 Ford cup at 25°C) – s	20
Suggested bake 20–15 min at 400°F	

Corrosion-Resistant Baking Primer

FORMULA NO. 1

(Alkyd)

	lb	gal
Grind in sand mill:		
M-50 Basic Lead Silicochromate	170.0	4.98
R-9098 Red Iron Oxide	170.0	4.80
Nytal 300 Talc	170.0	7.16
Celite 281 Diatomaceous Silica	70.0	3.64
Ben-A-Gel Thickener	2.0	0.13
Aquamac 1065	255.0	30.00
Propasol P Solvent	125.1	17.00
Let down:		
Aquamac 1065	110.5	13.00
Water	147.2	17.65
Ammonia (28%) to minimum pH of 8.5	4.2	0.55
Cymel 350	23.7	2.37

PVC (%)	42.7
Vehicle Nonvolatile (% by Wt.)	39.6
Wt./gal (lb)	12.33
Visc. (#4 Ford cup) – s	150
pH	8.5
Bake for 15 min at 350 F	

No. 2

(Alkyd)

	lb	gal
Grind in sand mill:		
Tribasic Lead Chromosilicate	102.0	2.04
Ti-Pure R-900 Titanium Dioxide	34.0	0.97
Barytes	155.0	4.12
Carbon Black	2.0	0.13
Mistron HGO-55 Talc	78.0	3.57
Ben-A-Gel Thickener	5.6	0.37
Aquamac 1065	102.0	12.00
Propasol P Solvent	110.4	15.00
Let down:		
Aquamac 1065	212.5	25.00
Ammonia (28%) to minimum pH of 9.0	4.0	0.53
Cymel 350	20.5	2.05
Water	308.4	37.00
PVC (%)	33.5	
Vehicle Nonvolatile (% Wt.)	29.6	
lb/gal	11.04	
Visc. (#4 Ford cup) − s	54	
pH	9.2	
Bake for 15 min at 350 F		

Heat-Resistant Primer

(Alkyd)

Roller mill grind:	
Titanium Dioxide	51.2
Duraplex ND-77B (60%)	36.4
High-Flash Naphtha	5.5
Mix with:	
Xylene	6.9
Total Solids (%)	73.0

Titanium Dioxide	70%
Duraplex ND-77B	30%

Visc. (approximate) (No. 4 Ford Cup) 90 s

Application:

For maximum gloss, the primer must be applied in a very thin coat. A spraying viscosity of 15 s or a solids of 55% is suggested. Best results are obtained by using a partially throttled gun and reduced air pressure. The primed surface is baked for 20 min at 400 F. Lower baking temperatures reduce the adhesion and gloss of the enamel. Air-drying of the primer is not necessary.

Due to the comparatively thin enamel coat to be applied, it is important that the primer be applied uniformly. Best results are obtained by sanding the primer coat to remove any surface irregularities. However, if the primer coat is uniform, sanding is not necessary.

Orange Baking Enamel

(Alkyd)

	lb	gal
Softex Ming Orange Dark 2522-03		
Molybdate Orange Pigment	200.0	6.00
Duramac 2483 Alkyd Resin (50%)	450.0	54.54
Cymel 248-8 Melamine Coating		
Resin (55%)	50.0	5.95
Plaskon 3353 Urea Coating Resin		
(50%)	90.0	10.71
Butanol	50.0	7.46
V M & P Naphtha	53.0	8.40
Xylene	36.0	4.93
Adamac Catalyst 20	14.7	2.01

Visc. (#4 Ford cup) − s	25
Wt./gal (lb)	9.44
Total Solids (% by Wt.)	52.65
Nonvolatile Vehicle	75.6% alkyd
	9.3% melamine
	15.1% urea
Catalyst Level:	5% on NVV

Baking Schedule to F pencil hardness:

| With **Catalyst 20** | 6 min at 300 F |
| Without Catalyst | 20 min at 300 F |

Properties on **Bonderite** steel panel baked 12 min at 250 F:

Film Thickness	0.8 mil
Pencil Hardness	HB
60° Gloss Reading	82
Impact, direct	20 in.–lb
reverse	8 in.–lb
Adhesion	97%
Flexibility on Mandrel	OK

Purple Baking Enamel

(Alkyd)

	lb	gal
Monastral Violet R (RT795-D)	4.0	0.30
MO-588 Aluminum Paste	8.0	0.60
Duramac 2483 Alkyd Resin (50%)	450.0	54.54
Cymel 248-8 Melamine Coating Resin (50%)	55.0	6.55
Uformite F-200E Urea Coating Resin (50%)	90.0	10.71
Butanol	50.0	7.46
V M & P Naphtha	58.0	9.20
Xylene	62.0	8.58
Adamac Catalyst 20	14.7	2.01

Visc. (#4 Ford cup) – s	29
Wt./gal (lb)	7.92
Total Solids (% by Wt.)	40
Nonvolatile Vehicle	75% alkyd
	10% melamine
	15% urea
Catalyst Level:	5% on NVV
Baking schedule	5 min at 300 F

Properties on **Bonderite** steel panel:

	lb	gal
Film Thickness	0.5 mil	
Pencil Hardness	F	
60° Gloss Reading	91	
Impact, direct	16 in.–lb	
reverse	<4 in.–lb	
Adhesion	98%	
Flexibility on Mandrel	OK	

Black Baking Enamel

FORMULA No. 1

(Alkyd)

	lb	gal
Peerless Mark II	25.0	1.65
Duramac 2483 Alkyd Resin (50%)	450.0	54.54
Cymel 248-8 Melamine Coating Resin (55%)	55.0	6.55
Plaskon 3353 Urea Coating Resin (50%)	90.0	10.71
Butanol	50.0	7.46
V M & P Naphtha	58.0	9.20
Xylene	56.0	7.81
Adamac Catalyst 20	14.7	2.01

Visc. (#4 Ford cup) – s	29
Wt./gal (lb)	7.91
Total Solids (% by Wt.)	39
Nonvolatile Vehicle	75% alkyd
	10% melamine
	15% urea
Catalyst Level:	5% on NVV
Baking Schedule	5 min at 300°F

Properties on **Bonderite** steel panel:

Film Thickness	0.5 mil
Pencil Hardness	F
60° Gloss Reading	94

INDUSTRIAL PAINTS 357

	lb	gal
Impact, direct	16 in.–lb	
reverse	<4 in.–lb	
Adhesion	98%	
Flexibility on Mandrel	OK	

No. 2

(Alkyd)

	lb	gal
Grind in sand mill:		
Carbon Black	11.0	0.73
Aquamac 1065	102.0	12.00
Let down:		
Aquamac 1065	246.5	29.00
Ammonia (28%) to minimum		
pH = 8.3	4.0	0.53
Water	458.2	55.0
Paint Additive #11 Silicone Resin	5.0	0.70

PVC (%)	3.0
Vehicle Nonvolatile (% by Wt.)	27.7
Total Solids (% by Wt.)	28.7
Wt./gal (lb)	8.45 lb
Visc. (#4 Ford cup) – s	117
pH	8.4
Bake for 15 min at 350 F	

Red Baking Enamel

(Alkyd)

	lb	gal
Grind in sand mill:		
Toluidine red	60.0	5.14
Aquamac 1065	102.0	12.00
Water	41.6	5.00

	lb		gal
Let down:			
Aquamac 1065	323.0		38.00
Ammonia (28%) to minimum			
pH — 8.3	6.0		0.80
Water	333.3		40.00
Paint Additive #11 Silicone Resin	5.0		0.70
PVC (%)		15.2	
Vehicle Nonvolatile (% by Wt.)		34.4	
Total Solids (% by Wt.)		38.6	
Wt./gal (lb)		8.58	
Visc. (#4 Ford cup) — s		193	
pH		8.5	
Bake for 15 min at 350 F			

Blue Baking Enamel

(Alkyd)

	lb		gal
Grind in sand mill:			
G-297-D Monastral Blue	40.0		3.08
Aquamac 1065	59.6		7.00
Isobutyl Alcohol	73.5		11.00
Let down:			
Aquamac 1065	298.2		35.00
Ammonia (28%) to minimum			
pH — 8.3	6.0		0.80
Water	383.2		46.00
Paint Additive #11 Silicone Resin	5.0		0.70
PVC (%)		10.2	
Vehicle Nonvolatile (% by Wt.)		30.6	
Total Solids (% by Wt.)		33.8	
Wt./gal (lb)		8.48	
Visc. (#4 Ford cup) — s			
Fresh		68	

4 Months   68
pH   8.5
Bake for 15 min at 350 F

Yellow Baking Enamel

(Alkyd)

	lb	gal
Hy-Cap Y-469 Chrome Yellow		
Pigment	250.0	5.60
Duramac 2483 Alkyd Resin (50%)	450.0	54.54
Uformite MM-55 Melamine Coating		
Resin (50%)	55.0	6.79
Beetle XB-1032 Urea Coating		
Resin (50%)	90.0	10.71
Butanol	50.0	7.46
V M & P Naphtha	53.0	8.40
Xylene	32.0	4.49
Adamac Catalyst 20	14.7	2.01

Visc. (#4 Ford cup) — s	28
Wt./gal (lb)	9.95
Total Solids (% by Wt.)	55.28
Nonvolatile Vehicle	75% alkyd
	10% melamine
	15% urea
Catalyst Level:	4.9% on NVV
Baking Schedule	10 min at 250 F
Range:	2 min at 350 F

Properties on **Bonderite** steel panel baked 10 min at 250 F:

Film Thickness	0.8 mil
Pencil Hardness	B
60° Gloss Reading	92
Impact, direct	12 in.–lb
reverse	<4 in.–lb
Adhesion	98%
Flexibility on Mandrel	OK

White Baking Enamel

FORMULA No. 1

(Alkyd)

	lb	gal
Ti-Pure R-900 Titanium Dioxide	250.0	7.30
Duramac 2483 Alkyd Resin (50%)	450.0	54.54
Plaskon 3353 Urea Coating		
Resin (50%)	120.0	14.55
Butanol	50.0	7.46
V M & P Naphtha	63.0	10.00
Xylene	38.0	5.28
Adamac Catalyst 20	8.5	1.17

Visc. (#4 Ford cup) − s	26
Wt./gal (lb)	9.79
Total Solids (% by Wt.)	54.62
Nonvolatile Vehicle	79% alkyd
	21% urea
Catalyst Level	3% of NVV

Baking schedule to F pencil hardness:

With **Catalyst 20**	6 min at 300°F
Without Catalyst	30 min at 300°F

Properties on **Bonderite** steel panel baked 15 min at 250°F

Film Thickness	0.8 mil
Pencil Hardness	B
60° Gloss Reading	84
Impact, direct	8 in.–lb
reverse	<4 in.–lb
Adhesion	98%
Flexibility on Mandrel	OK

No. 2

(Alkyd)

	lb	gal
Ti-Pure R-900 Titanium Dioxide	250.0	7.30
Duramac 2483 Alkyd Resin (50%)	450.0	54.54

	lb	gal
Uformite MM-55 Melamine Coating		
Resin (50%)	120.0	14.81
Butanol	50.0	7.46
V M & P Naphtha	63.0	10.00
Xylene	28.0	3.88
Adamac Catalyst 20	14.7	2.01

Visc. (#4 Ford cup) – s	22
Wt./gal (lb)	9.76
Total Solids (% by Wt.)	54.8
Nonvolatile Vehicle	79% alkyd
	21% melamine
Catalyst Level:	5.2% of NVV

Baking schedule to F pencil hardness:

With **Catalyst 20**	6 min at 300°F
Without Catalyst	20 min at 300°F

Properties on **Bonderite** steel panel baked 10 min at 250°F

Film Thickness	0.8 mil
Pencil Hardness	H
60° Gloss Reading	83
Impact, direct	16 in.–lb
reverse	<4 in.–lb
Adhesion	95%
Flexibility on Mandrel	OK

No. 3

(Alkyd)

	lb	gal
Ti-Pure R-900 Titanium Dioxide	250.0	7.30
Duramac 2483 Alkyd Resin (50%)	450.0	54.54
Cymel 248-8 Melamine Coating		
Resin (55%)	50.0	5.95
Plaskon 3353 Urea Coating		
Resin (50%)	90.0	10.71
Butanol	50.0	7.46
V M & P Naphtha	53.0	8.40

	lb	gal
Xylene	26.0	3.63
Adamac Catalyst 20	14.7	2.01

Visc. (#4 Ford cup) – s	27
Wt./gal (lb)	9.84
Total Solids (% by Wt.)	55.59
Nonvolatile Vehicle	75.6% alkyd
	9.3% melamine
	15.1% urea
Catalyst Level:	4.95% on NVV
Baking schedule	12 min at 250°F
range:	2 min at 350°F

Properties on **Bonderite** steel panel baked 12 min at 250°F

Film Thickness	0.5 mil
Pencil Hardness	2H
60° Gloss Reading	86
Impact, direct	16 in.–lb
reverse	4 in.–lb
Adhesion	98%
Flexibility on Mandrel	fails

No. 4

(Alkyd)

Rutile Titanium Dioxide	240
Barytes X-10R	240
Beckosol 1307-50HV	550
Post-4	8
Uformite 200E	50
Solvesso 100	126

White Kitchen Cabinet Enamel

(Alkyd)

	lb	gal
Roller mill grind:		
Rutile Titanium Dioxide	336	10.4
Duraplex ND-77B (60% solids)	302	35.2
Xylene	25	3.5
Mix with:		
Duraplex ND-77B (60% solids)	53	6.2
Uformite F-210 (50% solids)	182	21.5
Xylene	167	23.2
Wt./gal (lb)	10.7	
Total Solids (%)	60.0	
Pigment (%)	52.5	
Binder (%)	47.5	
Uformite F-210	30.0%	
Duraplex ND-77B	70.0%	

The above formulation is at storage viscosity at 60% solids. It can be reduced with xylene for spray application at about 50% solids.

A summary of the application characteristics of this enamel is given below.

Approximate visc. at 60% solids	
(No. 4 Ford Cup) – s	68
Spray Solids (20 s in No. 4 Ford Cup)	49.8%
Pencil hardness (30 min bakes)	
250°F	3B
300°F	F
Adhesion (30 min at 300 F.)	Good
Photovolt gloss	85.8%
Color (2 hr at 300°F.)	Excellent
NaOH resistance (1.5% NaOH – panels baked 30 min at 300°F.)	
1 hr	no effect
2½ hr	no effect

6½ hr	minute blistering
24 hr	tiny blisters — no dulling
48 hr	tiny blisters — very slight dulling

White Enamel for Appliances

(Alkyd)

	lb	gal
Roller mill grind:		
Titanium Dioxide	355	10.9
Duraplex ND-78 (60% solids)	236	27.8
·Mix with:		
Duraplex ND-78 (60% solids)	177	20.7
Uformite MX-61 (60% solids)	177	20.4
Xylol	146	20.2
Wt./gal (%)	10.9	
Total Solids (%)	65	
Pigment (%)	50	
Binder (%)	50	
Uformite MX-61	30%	
Duraplex ND-78	70%	

The enamel as made at 65% solids is at storage viscosity. It will spray satisfactorily at about 58% solids when reduced with xylol. The suggested baking schedule is 30 min at 300 F.

This enamel, when applied in a single coat over bonderized steel and baked for 30 min at 300 F., will pass a 100-hr soak test in a 0.5% **Rinso** solution at 165 F. It will withstand immersion in 30% sodium hydroxide for 9 days at room temperature and still retain excellent gloss. Even after a drastic overbake of 40 hr at 325 F., the film retains good color and high gloss.

For highest resistance to alkali, the alkyd portion of the vehicle should be comprised solely of **Duraplex ND-78**. However, **Duraplex A-27**, to the extent of 25% of the alkyd component, can be substituted to give slightly greater adhesion with only a slight sacrifice in resistance to soap and alkali. This latter type of formulation permits one-coat application.

Gray Corrosion-Resistant Baking Enamel

(Alkyd)

	lb	gal
Grind in pebble mill:		
Ti-Pure R-901 Titanium Dioxide	49.0	1.44
Black Iron Oxide	18.0	0.44
Tribasic Lead Chromosilicate	50.0	1.00
Aquamac 1065	80.0	9.42
Water	60.0	7.21
Ammonia (28%) to minimum		
pH — 9.0	2.0	0.26
Let down:		
Aquamac 1065	340.0	40.00
Ben-A-Gel Thickener	100.0	12.00
Water	200.0	24.00

PVC (%)	8.65	
Vehicle Nonvolatile (% by Wt.)	34.90	
Wt./gal (lb)	9.39	
Visc. (#4 Ford cup) — s	55	
pH	9.1	

Bake for 15 min at 385 F

Gloss Baking Enamel

FORMULA No. 1

(Alkyd)

	lb	gal
Varkyd 363-50J	165	20.0
Solvent 102 J	64	10.0
H F Thinner	50	7.0
M-P-A 60 (Xylene)	8	.9
Titanium Dioxide **Ti-Pure R-900**	265	7.9
Sand mill or pebble mill:		
Varkyd 363-50J	388	47.0

	lb	gal
Solvent 102J	64	9.0
Cobalt Naphthenate (6%)	.5	—
PVC (%)		21.7
Vehicle Nonvolatile (Wt.)		37.7
Total N.V. by Wt. (%)		54.2
Wt./gal (lb)		9.95
Visc. (KU)		60–65
Reduction to 20–25 s #4 Ford Cup with **Solvent 102J**		
Spraying Properties		Excellent
Leveling		Excellent
Sagging Resistance		Excellent
Settling		None
Skinning		None
Gloss		Excellent
Bake Schedule		20 min at 300°F
Hardness		Very Good
Marproofness		Good
Adhesion		Excellent
Flexibility on tin 1/8″ Bend		Excellent

No. 2

(Alkyd)

	lb	gal
Varkyd 1577-50 VMAL	152	20.0
M-P-A 60 (Xylene)	8	0.9
Butyl Cellosolve	30	4.0
Nuosperse 657	4	.5
H F Thinner Pacific Petrochem	77	12.0
TiO_2 **Ti-Pure R-900**	275	7.9

Sand grind or pebble mill:

	lb	gal
Solvent 102.1 Pacific Petrochem	21	3.0
Varkyd 1577 – Sovmal	365	48.0
Solvent 102.1	42.5	6.0
Cobalt Naphthenate (6%)	.5	—

PVC (%)	22.0
Vehicle Nonvolatile (% by Wt.)	37.8
Total N.V. (% by Wt.)	55.3
Wt./gal (lb)	9.53
Visc. (KU)	60–65
Reduction to 20–25 s #4 Ford Cup with solvent	102.1
Spraying Properties	Very Good
Leveling	Good–Very Good (V.Sl. Orange Peel)
Settling	Very Very Slight
Skinning	None
Gloss	Very Good
Bake Schedule	20 min at 300 F
Hardness	Excellent
Marproofness	Excellent
Flexibility 1/8" Bend	Excellent

No. 3

(Alkyd)

	lb	gal
Varkyd 1577-45 VMJ	266	35.0
M-P-A 60 (Xylene)	8	.9
Butyl Cellosolve	30	4.0
Nuosperse 657	4	.5
Solvent 102J	14	2.0
Titanium Dioxide **Ti-Pure R-900**	275	7.9

Sand grind or pebble mill:

	lb	gal
Varkyd 1577-45 VMJ	350	46.0
Melamine VN-812	56	6.5

PVC (%)	20.5
Vehicle Nonvolatile (% by Wt.)	42.2
Total N.V. (% by Wt.)	55.8
Wt. per gal (lb)	9.77
Visc. (KU)	60–65
Reduction to 20–25 s #4 Ford Cup with Solvent 102J	

Spraying Properties	Excellent
Leveling	Excellent
Settling	None
Skinning	None
Gloss	Excellent
Bake Schedule	20 min at 300 F
Hardness	Excellent
Marproofness	Very Good
Adhesion	Excellent
Flexibility on Tin 1/8″ Bend	Excellent

White Metal-Decorating Enamel

(Alkyd)

	lb	gal
Roller mill grind:		
Titanium Dioxide	278	8.6
Duraplex C-58 (60% solids)	210	26.8
Mix with:		
Duraplex C-58 (60% solids)	252	32.6
Mineral Thinner	189	29.1
Solvesso 150	21.2	2.8
Cobalt Naphthenate (6%)	0.9	0.1
Wt./gal (lb)		9.5
Total Solids (%)		58.5
Pigment (%)		50.0
Vehicle (%)		50.0
Thinner:		
Mineral Spirits		90.0%
Solvesso 150		10.0%
Drier (metal on alkyd solids)		0.02%

Bake for 10 min at 350 F.

Gray, High-Speed Baking Primer

(Alkyd)

Ti-Pure R-901 Titanium Dioxide	240.0	7.22
Lampblack	4.0	0.26

Barytes X5R	100.0	2.79
Aquamac 1065 Oxazoline Alkyd Resin (65%)	480.0	56.47
Cymel 350 Melamine Coating Resin (98%)	96.0	9.60
Butyl Cellosolve	35.0	4.68
Water	125.0	15.06
Aqua Ammonia (28%) to pH 8.60	13.0	1.73
Dow Corning DC-11 Antifoam	3.0	0.41
Adamac Catalyst 20	20.0	2.70

Visc. (#4 Ford cup) – s	115
Wt./gal	11.06
PVC (%)	19.54
Total Solids (% by vol.)	67.28
	51.67

Baking schedule	2 min at 450 F
range:	1 min at 500 F

Properties on **Bonderite** steel panel baked 2 min at 450 F:

Film Thickness	0.5 mil
Pencil Hardness	4H
60° Gloss Reading	42
Impact, direct	12 in.–lb
reverse	<4 in.–lb
Adhesion	100%
Flexibility on Mandrel	fails

Clear Baking Enamel

Formula No. 1

(Alkyd)

G-E75108 Methylon Resin	33.3
Diacetone Alcohol	20.0
Aroplaz 2480-XB-50 (50% resin sol'n.)	44.2
10% H_3PO_4 in Diacetone Alcohol	2.5 (0.45% H_3PO_4 on resin solids)
Resin Solids (%)	55.4

Visc. (#4 Ford cup at
   25 C) (s)   44

Suggested bake 15–20 min at 400 F

Resistant to solvents, chemicals, soap solutions, salt solutions, etc. Excellent flexibility. High resin solids content.

Excellent primer vehicle. The alkyd improves resistance to salt spray immediate to scribed area, allows higher pigment concentrations and lower cost. The 75108 greatly improves the chemical resistance. The humidity resistance is outstanding and the resistance to strong solvents is superior to that of straight alkyds.

No. 2

(Alkyd)

G-E 75108 Methylon Resin	17.0
Epoxy Resin Sol'n. (40%) in 1–1 Xylene-Diacetone Alcohol	43.4
Diacetone Alcohol	23.6
Xylene	2.5
Resimene 881 (60% resin sol'n.)	6.3
H_3PO_4 (10%) in Diacetone Alcohol	3.2 (0.85% H_3PO_4 on resin solids)
G-E SR-82 Silicone Resin (1%) in Xylene	5.0 (0.13% SR-82 on resin solids)
Resin Solids (%)	37.7
Visc. (#4 Ford cup at 25°C) – (s)	26

Suggested thinning solvents: equal parts xylene and diacetone alcohol

Suggested bake 12 min at 400 F
            40 min at 350 F

Excellent chemical, solvent, alkali, acid, soap and water resistance plus film flexibility. Suggested for use in drum and container linings, etc.

No. 3

(Epoxy Blend)

G-E 75108 Methylon Resin	18.2
Epoxy Resin Sol'n. (40%) in	
1–1 Xylene-Diacetone	
Alcohol	45.5
Diacetone Alcohol	18.2
Xylene	4.5
H_3PO_4 (10%) in Diacetone	
Alcohol	3.6 (1.0% H_3PO_4 on resin solids)
1–1–1–1 Diacetone Alcohol–	
Xylene–Denatured Alcohol–	
Toluene Sol'n.	4.5
SR-82 Silicone Resin (1%)	
in Xylene	5.5 (0.15% SR-82 on resin solids)
Resin Solids (%)	36.4%

Suggested bake 15–20 min at 365 F
12–15 min at 400 F

Multiple-coat system   Intermediate bake   5–6   min at 400 F
Final bake   30–40 min at 400 F

Using the above formulation, an unscribed 1½ mil coating (2-coat system) on plain, clean steel with edges protected showed no effect after exposure to salt spray for two years. The final bake was 20–25 min at 400 F.

Designed for maximum chemical, solvent, alkali and acid resistance plus some film flexibility.

No. 4

(Epoxy Blend)

G-E 75108 Methylon Resin	8.8
Epoxy Resin Sol'n. (40%) in	
1–1 Xylene–Diacetone	
Alcohol	65.7
1–1–1–1 Diacetone Alcohol–	
Xylene–Denatured Alcohol–	
Toluene Sol'n.	21.4

H₃PO₄ (10%) in Diacetone
   Alcohol   3.5 (1% H_3PO_4 on resin solids)
G-E SR-82 Silicone Resin
   (10%) in Xylene   0.6 (0.17% SR-82 on resin solids)

Resin Solids (%)   35.1

Suggested bake 15 min at 400 F

Excellent chemical, solvent, alkali and acid resistance. It is designed for use in drum and container corrosion-resistant linings where rough handling is expected.

No. 5

(Epoxy Blend)

G-E 75108 Methylon Resin	13.4
Epoxy Resin Sol'n. (40%) in 1-1 Xylene–Diacetone Alcohol	50.2
Diacetone Alcohol	16.2
H_3PO_4 (10%) in Diacetone Alcohol	3.4 (1.0% H_3PO_4 on resin solids)
1-1-1-1 Diacetone Alcohol–Xylene–Denatured Alcohol–Toluene Sol'n.	11.7
G-E SR-82 Silicone Resin (1%) in Xylene	5.1 (0.15% SR-82 on resin solids)
Resin Solids (%)	33.5
Visc. (#4 Ford cup at 25 C) – (s)	21

Suggested bake   15–20 min at 365 F
                    12–15 min at 400 F

Multiple-coat system Intermediate bake 5–6 min at 400 F
                     Final bake   30–40 min at 400 F

Designed for maximum chemical, solvent, alkali and acid resistance plus very good film flexibility.* Suggested for use in drum and container corrosion-resistant linings, where continuous rough handling is expected.

Versatile, with a favorable balance of chemical resistance and flexibility, it is an excellent starting formulation for any corrosion-resistant application. Shelf life —about one year.

---

*3–6 months resistance to 16% caustic at 195 F (90 C) is normal for 4–6 mil multiple coats. Comparable caustic resistance is also possible at baking temperatures as low as 300 F using 1–1.5% phenol sulfonic acid in place of phosphoric acid. The steel substrate must be wash primed however to prevent attack by the acid. The acetone resistance is lessened with phenol sulfonic acid; the coating can be scratched with a thumbnail after 45 min immersion in acetone.

Gray Epoxy-Urea Flexible Baking Finish

	lb/100 gal
Titanox RA	53.0
Superjet Lampblack	6.3
RCl 45-410 Ferndale Chrome Green	2.0
Nytal 300	53.0
Imperial IAF Compound	2.0
Epotuf 38-503	324.0
Beckamine 21-511	74.0
Epotuf 37-151	29.0
Cellosolve	175.0
Xylol	175.0

Steel ball mill to 7 N.S. fineness

Methyl Paratoluene Sulfonate — 20% sol'n. in Ethyl Alcohol	4.5

Uses:

Metal appliance finish where outstanding flexibility and adhesion are required.

Epoxy — Polyurethane Finish Coat

Epoxy Mill Base:

Epon 1009 (or equivalent)	199
Half Second Butyrate (10% in ethylglycol acetate)	7
Titanium Dioxide (**R-KB-2**)	175

Ethyl Glycol Acetate	416
Desmodur Sol'n.:	
Desmodur N-75	203
Solids (%)	53
NCO/OH	1.0
Pigment/Binder	0.5/1

Reduce to spray with the following solvent mixture: methyl ethyl ketone, butyl acetate, ethylglycol acetate, toluene (4:1:4:1 pbw).

To accelerate the cure rate, 0.2% zinc octoate (8% Zn) or 0.5% **Desmorapid PP** (each based on resin solids) may be added.

The addition of 1% Bentone 27, based on pigment, reduces the hard settling tendencies. This finish coat is noted for high solvent and chemical resistance as well as resistance to color change and chalking in exterior exposures. For the ultimate in weather resistance, this finish should be top coated with a **Desmophen 650A/Multron R-221/Desmodur N-75** system.

Unpigmented Baking Finish

(Epoxy)

	lb	gal
Araldite 485 E-50	470.5	52.88
Cellosolve Acetate	18.2	2.24
Toluene	183.4	25.29
Urea-Formaldehyde Resin		
(60% N.V.)	168.0	19.59
Nonvolatile Content (%)	40	
Epoxy/UF Resin Ratio (solids)	70/30	

Cure Schedule:	30 min at 177 C (350 F)
Substrate	Steel
Film Thickness (mil)	0.7
Impact Resistance, reverse, in.–lb.	40
Flexibility, Conical Mandrel	Pass
MEK Resistance, 5 hr	No effect
37% HCl Resistance, 30 min	Blistering
Boiling 20% NaOH Resistance, 90 min	Loss of adhesion

Clear Chemical-Resistant Epoxy-Phenolic Baking Finish

	lb/100 gal
Epotuf 38-504	418.0
Ethyl Alcohol	93.0
Methyl Ethyl Ketone	37.0
Pine Oil	32.0
No. 840 Silicone Resin	3.0
Varcum 29-160	157.0
Toluol	42.0
Ortho Phosphoric Acid (85%)	5.8
Diacetone Alcohol } Age Overnight*	23.5
Butyl Alcohol	10.2

*The phosphoric acid should be added to the diacetone alcohol and butyl alcohol and allowed to age overnight before adding to the formula. This formulation can be pigmented by grinding in the epoxy resin.

General Purpose Epoxy Flash-Dry Primer

Mill Base (ball mill, 120 F):

Red Lead (97%)	86.6
Red Iron Oxide	86.6
Talc	74.1
Grinding Aid	7.4
Dispersing Aid	2.4
Eponol 55-B-40	253.0
CKR-5254	11.7
Acetone	97.2
Toluene	107.0
Pentoxone	24.7
M-P-A 60 (Toluene)	70.1

Mill Rinse:

Acetone	47.1
Toluene	51.7
Pentoxone	12.0

Application:

Thin one volume of primer base with one volume of spray thinner. The presence of **M-P-A 60** (Toluene) (a) overcame the pronounced settling of pigment in the thinned formula on prolonged standing, (b) improved the spray characteristics, and (c) brought the sagging tendency under control.

(Composition: 45 acetone/45 toluene/10 pentoxone).

Epoxy Gloss Enamel

(Amine Converted)

	lb	gal
Epoxy Resin	213.8	22.38
Thixatrol ST	4	0.472
Rutile Titanium Dioxide	177.2	5.06
Chrome Oxide	36.7	0.85
Cellosolve Acetate	156	19.25
Xylene	156	21.20
Beckamine P-196	152.7	17.81
Cellosolve Acetate	50	6.18
Xylene	50	6.80

Procedure:

Prepare a solution of **Epon 1007** in **Cellosolve** acetate/xylene (1/1 by weight) at a suitable solids concentration so that all of the Epon resin is contained in the grind. Disperse the pigment and **Thixatrol ST** in this solution in a pebble mill (at about 100 F). Let down with the **Beckamine P-196** and remaining solvents. Poor gloss may result if part of the **Epon** resin is used in letting down this dispersion.

Application:

For spray application reduce further with **Cellosolve** acetate/toluene (1/1) by weight. Recommended baking schedule: 20 min at 385 F.

White Fast Baking Epoxy-Urea Enamel

	lb/100 gal
Titanox RA	300.0
Beckamine 21-511	107.0
Toluol	57.0

Pebble mill to 7 + N.S. fineness

	lb/100 gal
Beckamine 21-511 (P-196-60)	55.0
Toluol	129.0
Epotuf 38-503 (6503-55)	414.0
Methyl Paratoluene Sulfonate — 20% Sol'n.	
in Ethyl Alcohol	9.7

Uses:

Highly resistant application finish, laboratory and hospital metal furniture finish.

General Purpose Epoxy Flash-Dry Primer

	lb	gal
Red Lead (97%)	94.9	1.26
Red Iron Oxide (**Mapico 516**)	94.9	2.21
Asbestine 3X	81.1	3.41
Nuosperse 657	2.7	.34
Eponol 53-L-32	346.2	39.89
M-P-A 60 (Xylene)	76.6	10.52
Roller mill grind:		
CKR5254 Sol'n. (50%)	60.4	7.25
Methyl Cellosolve	119.8	14.93
Cyclo Sol 53	66.6	9.16
Xylene	79.9	11.03

Procedure:

The pigment, extender, **Nuosperse 657**, **Eponol 53-L-32**, and **M-P-A 60** are dispersed on a 3 roll mill. This dispersion is then let down with the phenolic resin solution and the remaining solvent.

Epoxy Flat White Baking Enamel

Formula No. 1

	lb	gal
Titanium Dioxide **R-900**	250	7.15
Varkyd 522-60 HS	176	22.0

	lb	gal
Amsco Solv. B (Xylol Substitute)	85	12.0
M-P-A 60 (Mineral Spirits)	3	.4

Pebble Mill to 7 + Grind:

	lb	gal
Varkyd 522-60 HS	236	17.0
Resimene 879-50	126	15.0
Varkyd 606-60X	106	12.5
Amsco Solv. B	53	7.5
Amsco High-Flash Naphtha	58	8.0
Butyl **Cellosolve**	15	2.0

Application:

This enamel will provide very good chemical resistance, and its lack of gloss is extremely important in those applications demanding glare-free surfaces.

No. 2

	lb	gal
Titanium Dioxide **R-900**	250	7.15
Varkyd 522-60 HS	176	22.0
Amsco Solv. B (Xylol Substitute)	85	12.0
MPA-60	3	.4

Pebble mill to 7 + grind

	lb	gal
Varkyd 522-60 HS	136	17.0
Resimene 879-50	126	15.0
Varkyd 606-60X	106	12.5
Amsco Solv. B	53	7.5
Amsco Hi-Flash Naphtha	58	8.0
Butyl **Cellosolve**	15	2.0

PVC (%)	17.7
Vehicle Nonvolatile (% by Wt.)	41.7
Total NV by Wt.	56.0
Wt./gal	9.75
Visc. #4 Ford Cup − s	35–45
Reduction	Slight

Settling	None
Skinning	None
Gloss 60°	4–8 (Flat)
Baking Schedule	15 min at 300 F
Reflectance	80–85 Units
Hardness	Excellent
Toughness	Excellent
Mar Proofness	Very Good +
Adhesion	Excellent
Flexibility	Excellent

This enamel will provide very good chemical resistance, and its lack of gloss is extremely important in those applications demanding glare-free surfaces.

Clear Fast Baking Epoxy-Urea Finish for Brass

	lb/100 gal
Epotuf 38-504	333.0
Beckamine 21-511	202.0
Methyl Isobutyl Ketone	88.0
Methyl Ethyl Ketone	88.0
Xylol	95.0
Methyl Paratoluene Sulfonate – 20% Sol'n. in Ethyl Alcohol	6.0
Tinapal SFG – 1% Sol'n. in Methyl Isobutyl Ketone	0.6
No. 840 Silicone Resin	2.0
Paint Additive No. 11	1.0

Use:

Clear coating for polished metal surfaces.

Epoxy Water-Soluble Maintenance Enamels

FORMULA	No. 1	No. 2
	(Gray)	*(Light Green)*
TMA Resin 408 (65% NVM)	330	330
NH$_4$OH (28%)	15	15

	No. 1 (cont'd)	No. 2 (cont'd)
	(Gray)	*(Light Green)*
Water	410	410
Manganese Naphthenate (6%)	1	1
Cobalt Naphthenate (6%)	3	3
Antiskinning Agent	1	1
Titanium Dioxide[1]	100	90
Chromium Oxide Green[2]	−	25
Lamp Black[3]	3	−
Calcium Carbonate[4]	80	80
Basic Lead Silico Chromate[5]	100	100
Visc. #4 Ford Cup − s	30–35	30–35
pH	8.0–8.5	8.0–8.5
NVM (Wt. %)	47	48
PVC (%)	29	29

[1] R-610
[2] G-6099
[3] 8452-90
[4] Duramite
[5] M-50

One-Component, Stain-Resistant Baked Enamel

(Polyurethane)

Multron R-351-65	386
Titanium Dioxide **R-KB-2**	232
Modaflow (10% in ethylglycol acetate)	7
EAB 381-1/10 (10% in ethylglycol acetate)	33
Ethyl Glycol Acetate	116
Mondur HCB	226
% Solids	58
NCO/OH	1.05
PVC (%)	15

Curing Schedules:
60–90 s   @ 500 F
       5 min @ 400 F

10–15 min @ 350 F
30 min @ 300 F

This coating system is characterized by high resistance to household stains and chemicals. It is recommended for evaluation as an appliance enamel, where a high degree of flexibility is not required and the slight color change which occurs upon exposure to sunlight can be tolerated.

Thermosetting Acrylic Appliance Finish

Ti-Pure R-960 Titanium Dioxide	295.0	8.97
Acryloid AT-56 Thermosetting Acrylic Resin (50%)	450.0	55.69
Uformite MM-55 Melamine Coating Resin (50%)	120.0	14.25
Butanol	50.0	7.46
Cellosolve Acetate	26.0	3.29
Xylene	61.0	8.39
Adamac Catalyst 20	14.3	1.95
Wt./gal (lb)	10.16	
Total Solids (% by Wt.)	57	

Baking schedule to H pencil hardness:
With Catalyst 20 — 6 min at 300 F
Without Catalyst — 20 min at 300 F

Properties on Bonderite steel panel baked 15 min at 300 F:
Film Thickness — 0.7 mil
Pencil Hardness — 3H
60° Gloss Reading — 83
Impact, direct — 16 in.–lb
reverse — <4 in.–lb
Adhesion — 100%
Flexibility on Mandrel — fails

Uformite Resins with Acrylic-Vinyl Mixtures

Clear coatings and white enamels for metals made with a combination of Acryloid B-72 and Bakelite vinyl resin VAGH modified with Uformite F-240 and Paraplex G-60 show exceptional retention, gloss, heat stability, adhesion, and chemical resistance.

A suggested formulation for a clear coating of this type follows. When baked for 15 min at 325 F., this coating develops good hardness and partially thermosets. It gives excellent resistance to chemicals and shows high gloss and clarity.

Clear Coating for Metal

(Acrylic — Vinyl)

Combine:

Acryloid B-72 (40% solids)	175.0
Bakelite Vinyl Resin VAGH (20% solids in methyl isobutyl ketone)	75.0
Uformite F-240 (60% solids)	16.6
Paraplex G-60	5.0
Xylene	128.4
Total Solids (%)	25
Visc. (cps)	65
Composition (solids):	
Acryloid B-72	70%
Bakelite Vinyl Resin VAGH	15%
Uformite F-240	10%
Paraplex G-60	5%

Uformite Resins with Thermosetting Acrylics

Uformite nitrogen resins are used to crosslink hydroxyl functional thermosetting acrylic resins of the Acryloid series to produce a baking-type coating. The following formulation is a good starting point for evaluating this coating system.

White Baking Enamel

FORMULA No. 1

(Acrylic)

	lb	gal
Roller mill grind:		
Rutile Titanium Dioxide	230.0	6.82
Uformite MM-47	173.0	20.40

	lb		gal
and then add:			
Acryloid AT-56	484		59.00
Xylene	49		6.75
Solvesso 100	24		3.29
Butanol	24		3.66
Total Solids (%)		57.9	
Visc. (No. 4 Ford Cup) – s		60	
Pigment (%)		40.0	
Binder (%)		60.0	
Acryloid AT-56		70.0%	
Uformite MM-47		30.0%	

Raybo 3 (0.05% based on acrylic solids) should be added to the enamel. For spray application, reduce the base enamel with a solvent blend of xylene/**Solvesso 100**/n-butanol (2/1/1) to a No. 4 Ford-Cup viscosity of 17 s. Spray solids will be approximately 46%.

The following properties were obtained on films baked for 15 min at 275 F.

Tukon Hardness (KHN)[1]	12.3
Tukon Hardness (KHN)[2]	10.9
Pencil Hardness[1]	B
Xylene Resistance (15 min)[1]	6B
Photovolt Gloss[2]– 20°	77
– 60°	92
Mandrel Flexibility[3]	Pass ¼″

Substrates used to determine the above results are glass[1], primed **Bonderite 100,**[2] and unprimed **Bonderite 100.**[3]

No. 2

(Acrylic)

	lb		gal
Roller mill grind:			
Titanium Dioxide	253.0		7.27

	lb	gal
Acryloid AT-50 (50% solids)	253.0	30.48

Mix with:

	lb	gal
Acryloid AT-50 (50% solids)	364.1	43.87
Xylol	96.9	13.46
Solvesso 150	25.0	3.37
Cellosolve Acetate	27.2	3.36
Raybo 3 (antisilk agent)	0.5	0.07

Wt./gal (lb)	10.2
Total Solids (%)	55.3
Pigment (%)	45.0
Vehicle (%)	55.0

No. 3

(Acrylic)

Roller mill grind:

	lb	gal
Rutile Titanium Dioxide	230.0	6.82
Uformite MM-47	173.0	20.40

and then add:

	lb	gal
Acryloid AT-56	484	59.00
Xylol	49	6.75
Solvesso 100	24	3.29
Butanol	24	3.66

Total Solids (%)	57
Visc. (#4 Ford Cup) − s	60
Pigment (%)	40.0
Binder (%)	60.0
Acryloid AT-56	70.0%
Uformite MM-47	30.0%

Raybo 3 (0.05%) should be added to the enamel, based on acrylic solids. For spray application, reduce the base enamel with a solvent blend of xylol/**Solvesso 100**/n-butanol:2/1/1 to 17 s (#4 Ford Cup visc.).

Heat-Resistant Metal Topcoat

(Acrylic)

Roller mill grind:

Titanium Dioxide	15.2
Acryloid A-10 (30%)	7.6
Toluene	2.3

Mix with:

Acryloid A-10 (30%)	68.1
Monoplex DBS (dibutyl sebacate)	0.8
Diethyl Phthalate	0.8
Toluene	5.2
Total Solids (%)	37.9
Titanium Dioxide	40%
Acryloid A-10	60%
Approximate viscosity (No. 4 Ford Cup)	90 s

Application:

This enamel should be sprayed over the primer shown below. Optimum spraying viscosity is about 15 s (No. 4 Ford Cup) at 27% solids. Higher solids give too thick a coat and produce blistering on baking. Best spray results are obtained by using a partially throttled gun and reduced air pressure.

Two enamel coats are applied with a 1-hr air dry between coats in a dustless atmosphere. A ½ hr air dry is allowed before baking. Force-drying is not recommended due to the tendency to produce inferior gloss and a pebbly surface pattern. Prolonged drying before baking will give reduced gloss.

Baking at high temperature is required to obtain maximum hardness. Good results are obtained at 350 F for 30 min or at 400 F for 20 min.

White Baking Enamel

FORMULA No. 1

(Acrylic – Epoxy)

	lb	gal
Roller mill grind:		
Rutile Titanium Dioxide (**Titanox RA** or **Ti-Pure R-900**)	274.0	7.81
Acryloid AT-70 (50% solids) (add toners as desired)	182.6	22.00
Mix with:		
Acryloid AT-70 (50% solids)	218.4	26.23
Epon 1001 (50% solids in **Cellosolve** acetate)	267.0	29.70
Xylene	77.8	10.96
Cellosolve Acetate	26.3	3.29
Raybo 3 (antisilk agent)	0.6	0.08

Wt./gal (lb)	10.5
Total Solids (%)	58.0
Visc. (#4 Ford Cup) – s	60
Spray Solids	48.7
Spray Visc. #4 Ford Cup – s	21
Pigment (%)	45
Vehicle (%)	55
Acryloid AT-70	60%
Epon 1001	40%

Application Properties:

The application properties of **Acryloid AT-70** were tested in our laboratory using the white baking enamel in this formula. The enamel was reduced with a mixture of xylol and **Cellosolve** acetate in a 3/1 ratio to spray viscosity (19 s to 21 s in a #4 Ford Cup). The recommended minimum baking schedule of 30 min at 350 F was used.

No. 2

(Acrylic — Epoxy)

	lb	gal
Roller mill grind:		
Titanium Dioxide	259.5	7.45
Acryloid AT-51 (50% solids)	259.5	31.10
(Add toners as desired)		
Mix with:		
Acryloid AT-51 (50% solids)	279.2	34.30
Epoxy Resin (50% in		
Cellosolve acetate)*	95.5	10.65
Xylol	81.5	11.26
Solvesso 150	21.5	2.90
Cellosolve acetate	23.2	2.86
Raybo 3 (antisilk agent)	0.5	0.07
Wt./gal (lb)	10.2	
Total Solids (%)	56.6	
Pigment (%)	45.0	
Vehicle (%)	55.0	
Acryloid AT-51	85.0%	
Epoxy Resin	15.0%	

---

*Epon 1001 or other compatible epoxy resins.

White Baking Enamel for Metal Decorating

(Acrylic — Epoxy)

	lb	gal
Roller mill grind:		
Rutile Titanium Dioxide	339.9	9.76
Acryloid AT-75 (50% solids)	226.4	26.98
Mix with:		
Acryloid AT-75 (50% solids)	18.2	2.16
Acryloid AT-71 (50% solids)	244.8	29.14

	lb	gal
Epon 1001 (50% Cellosolve		
acetate)	66.8	7.45
Solvesso 150	139.5	18.77
Cellosolve Acetate	46.4	5.73
Raybo 3	0.5	0.07
Wt./gal (lb)	10.8	
Total Solids (%)	57.1	
Visc. (#4 Ford Cup) – s	82	
Pigment (%)	55	
Vehicle (%)	45	
Acryloid AT-75	44%	
Acryloid AT-71	44%	
Epon 1001	12%	

The application properties of this formulation and the preceding two were tested in our laboratory. The baking schedules of 10 min at 375 F. and 10 min at 400 F. were both used in evaluating these formulations. The latter baking schedule gives complete cure.

Acrylic General-Purpose Enamel

Acryloid A-21 is another hard acrylic resin which is useful in appliance finishing. The hard acrylics are sometimes formulated with plasticizers to improve the flexibility of the coating. Plasticized and unplasticized enamels follow together with an extensive evaluation of comparative properties determined at several bake schedules. The improvement in adhesion, hardness and gloss at higher bake schedules should be noted.

	FORMULA No. 1	No. 2
	(Plasticized Enamel)	*(Unplasticized Enamel)*
Roller mill grind:		
Titanium Dioxide (**Titanox RA**)	45.0	45.0
Acryloid A-21 (30%)	45.0	45.0
Cellosolve Acetate	10.0	10.0

White High-Gloss Water-Reducible Industrial Baking Enamel
(Polyester)

	lb/100 gal
Unitante OR-650 Titanium Dioxide	447.3
Cyplex 1600 Resin (75%)	357.9
Dimethylaminoethanol	18.0
Silicone Resin **L5310** (10%)	17.1
XC-4011 Resin (75%)	11.4

Disperse on 3-roll mill and add:

Cyplex 1600 Resin (75%)	238.0
Cymel 303 Hexamethoxymethylmelamine (100%)	111.0
Catalyst 1010	5.7
Dimethylaminoethanol	11.7

APPENDIX

*p*H Values

Acids	*p*H Value	Bases	*p*H Value
Hydrochloric Acid	1.0	Sodium Bicarbonate	8.4
Sulfuric Acid	1.2	Borax	9.2
Phosphoric Acid	1.5	Ammonia	11.1
Sulfuric Acid	1.5	Sodium Carbonate	11.6
Acetic Acid	2.9	Trisodium Phosphate	12.0
Alum	3.2	Sodium Metasilicate	12.2
Carbonic Acid	3.8	Lime, Saturated	12.3
Boric Acid	5.2	Sodium Hydroxide	13.0

*p*H Ranges of Common Indicators

	Useful *p*H Range
Thymol Blue	1.2 — 2.8
Bromphenol Green	2.8 — 4.6
Methyl Orange	3.1 — 4.4
Bromcresol Green	4.0 — 5.6
Methyl Red	4.4 — 6.0
Propyl Red	4.8 — 6.4
Bromcresol Purple	5.2 — 6.8
Brom Thymol Blue	6.0 — 7.6
Phenol Red	6.8 — 8.4
Litmus	7.2 — 8.8
Cresol Red	7.2 — 8.8
Cresolphthalein	8.2 — 9.8
Phenolphthalein	8.6 — 10.2
Nitro Yellow	10.0 — 11.6
Alizarin Yellow R	10.1 — 12.1
Sulfo Orange	11.2 — 12.6

International Atomic Weights

	Symbol	Atomic Number	Atomic Weight		Symbol	Atomic Number	Atomic Weight
Actinium	Ac	89	227	Neodymium	ND	60	144.27
Aluminum	Al	13	26.98	Neptunium	Np	93	237.00
Americium	Am	95	243	Neon	Ne	10	20.183
Antimony	Sb	51	121.76	Nickel	Ni	28	58.71
Argon	A	18	39.944	Niobium	Nb	41	92.91
Arsenic	As	33	74.91	Nitrogen	N	7	14.008
Astatine	At	85	210	Osmium	Os	76	190.20
Barium	Ba	56	137.36	Oxygen	O	8	16
Berkelium	Bk	97	249	Palladium	Pd	46	106.4
Beryllium	Be	4	9.013	Phosphorus	P	15	30.975
Bismuth	Bi	83	209.00	Platinum	Pt	78	195.09
Boron	B	5	10.82	Plutonium	Pu	94	242.00
Bromine	Br	35	79.916	Potassium	K	19	39.100
Cadmium	Cd	48	112.41	Praseodymium	Pr	59	140.92
Calcium	Ca	20	40.08	Promethium	Pm	61	145
Californium	Cf	98	249.00	Protactinium	Pa	91	231
Carbon	C	6	12.011	Radium	Ra	88	226.05
Cerium	Ce	58	140.12	Radon	Rn	86	222
Cesium	Cs	55	132.91	Rhenium	Re	75	186.22

International Atomic Weights (cont.)

	Symbol	Atomic Number	Atomic Weight		Symbol	Atomic Number	Atomic Weight
Chlorine	Cl	17	35.457	Rhodium	Rh	45	102.91
Chromium	Cr	24	52.01	Rubidium	Rb	37	85.48
Cobalt	Co	27	58.94	Ruthenium	Ru	44	101.10
Copper	Cu	29	63.54	Samarium	Sm	62	150.35
Curium	Cm	96	245	Scandium	Sc	21	44.96
Dysprosium	Dy	66	162.51	Selenium	Se	34	78.96
Erbium	Er	68	167.27	Silicon	Si	14	28.09
Europium	Eu	63	152	Silver	Ag	47	107.880
Fluorine	F	9	19	Sodium	Na	11	22.991
Francium	Fr	87	223	Strontium	Sr	38	87.63
Gadolinium	Gd	64	157.26	Sulfur	S	16	32.066
Gallium	Ga	31	69.72	Tantalum	Ta	73	180.95
Germanium	Ge	32	72.60	Technetium	Tc	43	99
Gold	Au	79	197.00	Tellurium	Te	52	127.61
Hafnium	Hf	72	178.50	Terbium	Tb	65	158.93
Helium	He	2	4.003	Thallium	Tl	81	204.39
Holmium	Ho	67	164.94	Thorium	Th	90	232.05
Hydrogen	H	1	1.0080	Thulium	Tm	69	168.94
Indium	In	49	114.82	Tin	Sn	50	118.70

International Atomic Weights (cont.)

	Symbol	Atomic Number	Atomic Weight		Symbol	Atomic Number	Atomic Weight
Iodine	I	53	126.91	Titanium	Ti	22	47.90
Iridium	Ir	77	192.20	Tungsten	W	74	183.86
Iron	Fe	26	55.85	Uranium	U	92	238.07
Krypton	Kr	36	83.80	Vanadium	V	23	50.95
Lanthanum	La	57	138.92	Xenon	Xe	54	131.30
Lead	Pb	82	207.21	Ytterbium	Yb	70	173.04
Lithium	Li	3	6.940	Yttrium	Y	39	88.92
Lutetium	Lu	71	174.99	Zinc	Zn	30	65.38
Magnesium	Mg	12	24.32	Zirconium	Zr	40	91.22
Manganese	Mn	25	54.94				
Mendelevium	Mv	101	256.00				
Mercury	Hg	80	200.61				
Molybdenum	Mo	42	95.95				

Temperature Conversion Tables

F	C	F	C	F	C	F	C	F	C
-40	-40.0	9	-12.8	58	14.4	107	41.7	156	68.9
-39	-39.4	10	-12.2	59	15.0	108	42.2	157	69.4
-38	-38.9	11	-11.7	60	15.6	109	42.8	158	70.0
-37	-38.3	12	-11.1	61	16.1	110	43.3	159	70.6
-36	-37.8	13	-10.6	62	16.7	111	43.9	160	71.1
-35	-37.2	14	-10.0	63	17.2	112	44.4	161	71.7
-34	-36.7	15	-9.4	64	17.8	113	45.0	162	72.2
-33	-36.1	16	-8.9	65	18.3	114	45.6	163	72.8
-32	-35.6	17	-8.3	66	18.9	115	46.1	164	73.3
-31	-35.0	18	-7.8	67	19.4	116	46.7	165	73.9
-30	-34.4	19	-7.2	68	20.0	117	47.2	166	74.4
-29	-33.9	20	-6.7	69	20.6	118	47.8	167	75.0
-28	-33.3	21	-6.1	70	21.1	119	48.3	168	75.6
-27	-32.8	22	-5.6	71	21.7	120	48.9	169	76.1
-26	-32.2	23	-5.0	72	22.2	121	49.4	170	76.7
-25	-31.7	24	-4.4	73	22.8	122	50.0	171	77.2
-24	-31.1	25	-3.9	74	23.3	123	50.6	172	77.8
-23	-30.6	26	-3.3	75	23.9	124	51.1	173	78.3
-22	-30.0	27	-2.8	76	24.4	125	51.7	174	78.9
-21	-29.4	28	-2.2	77	25.0	126	52.2	175	79.4
-20	-28.9	29	-1.7	78	25.6	127	52.8	176	80.0
-19	-28.3	30	-1.1	79	26.1	128	53.3	177	80.6
-18	-27.8	31	-0.6	80	26.7	129	53.9	178	81.1
-17	-27.2	32	0.0	81	27.2	130	54.4	179	81.7
-16	-26.7	33	0.6	82	27.8	131	55.0	180	82.2
-15	-26.1	34	1.1	83	28.3	132	55.6	181	82.8
-14	-25.6	35	1.7	84	28.9	133	56.1	182	83.3
-13	-25.0	36	2.2	85	29.4	134	56.7	183	83.9
-12	-24.4	37	2.8	86	30.0	135	57.2	184	84.4
-11	-23.9	38	3.3	87	30.6	136	57.8	185	85.0
-10	-23.3	39	3.9	88	31.1	137	58.3	186	85.6
-9	-22.8	40	4.4	89	31.7	138	58.9	187	86.1
-8	-22.2	41	5.0	90	32.2	139	59.4	188	86.7
-7	-21.7	42	5.6	91	32.8	140	60.0	189	87.2
-6	-21.1	43	6.1	92	33.3	141	60.6	190	87.8

Temperature Conversion Tables (cont.)

F	C	F	C	F	C	F	C	F	C
−5	−20.6	44	6.7	93	33.9	142	61.1	191	88.3
−4	−20.0	45	7.2	94	34.4	143	61.7	192	88.9
−3	−19.4	46	7.8	95	35.0	144	62.2	193	89.4
−2	−18.9	47	8.3	96	35.6	145	62.8	194	90.0
−1	−18.3	48	8.9	97	36.1	146	63.3	195	90.6
0	−17.8	49	9.4	98	36.7	147	63.9	196	91.1
1	−17.2	50	10.0	99	37.2	148	64.4	197	91.7
2	−16.7	51	10.6	100	37.8	149	65.0	198	92.2
3	−16.1	52	11.1	101	38.3	150	65.6	199	92.8
4	−15.6	53	11.7	102	38.9	151	66.1	200	93.3
5	−15.0	54	12.2	103	39.4	152	66.7	201	93.9
6	−14.4	55	12.8	104	40.0	153	67.2	202	94.4
7	−13.9	56	13.3	105	40.6	154	67.8	203	95.0
8	−13.3	57	13.9	106	41.1	155	68.3	204	95.6
205	96.1	254	123.3	303	150.6	352	177.8	401	205.0
206	96.7	255	123.9	304	151.1	353	178.3	402	205.6
207	97.2	256	124.4	305	151.7	354	178.9	403	206.1
208	97.8	257	125.0	306	152.2	355	179.4	404	206.7
209	98.3	258	125.6	307	152.8	356	180.0	405	207.2
210	98.9	259	126.1	308	153.3	357	180.6	406	207.8
211	99.4	260	126.7	309	153.9	358	181.1	407	208.3
212	100.0	261	127.2	310	154.4	359	181.7	408	208.9
213	100.6	262	127.8	311	155.0	360	182.2	409	209.4
214	101.1	263	128.3	312	155.6	361	182.8	410	210.0
215	101.7	264	128.9	313	156.1	362	183.3	411	210.6
216	102.2	265	129.4	314	156.7	363	183.9	412	211.1
217	102.8	266	130.0	315	157.2	364	184.4	413	211.7
218	103.3	267	130.6	316	157.8	365	185.0	414	212.2
219	103.9	268	131.1	317	158.3	366	185.6	415	212.8
220	104.4	269	131.7	318	158.9	367	186.1	416	213.3
221	105.0	270	132.2	319	159.4	368	186.7	417	213.9
222	105.6	271	132.8	320	160.0	369	187.2	418	214.4
223	106.1	272	133.3	321	160.6	370	187.8	419	215.0
224	106.7	273	133.9	322	161.1	371	188.3	420	215.6
225	107.2	274	134.4	323	161.7	372	188.9	421	216.1

Temperature Conversion Tables (cont.)

F	C	F	C	F	C	F	C	F	C
226	107.8	275	135.0	324	162.2	373	189.4	422	216.7
227	108.3	276	135.6	325	162.8	374	190.0	423	217.2
228	108.9	277	136.1	326	163.3	375	190.6	424	217.8
229	109.4	278	136.7	327	163.9	376	191.1	425	218.3
230	110.0	279	137.2	328	164.4	377	191.7	426	218.9
231	110.6	280	137.8	329	165.0	378	192.2	427	219.4
232	111.1	281	138.3	330	165.6	379	192.8	428	220.0
233	111.7	282	138.9	331	166.1	380	193.3	429	220.6
234	112.2	283	139.4	332	166.7	381	193.9	430	221.1
235	112.8	284	140.0	333	167.2	382	194.4	431	221.7
236	113.3	285	140.6	334	167.8	383	195.0	432	222.2
237	113.9	286	141.1	335	168.3	384	195.6	433	222.8
238	114.4	287	141.7	336	168.9	385	196.1	434	223.3
239	115.0	288	142.2	337	169.4	386	196.7	435	223.9
240	115.6	289	142.8	338	170.0	387	197.2	436	224.4
241	116.1	290	143.3	339	170.6	388	197.8	437	225.0
242	116.7	291	143.9	340	171.1	389	198.3	438	225.6
243	117.2	292	144.4	341	171.7	390	198.9	439	226.1
244	117.8	293	145.0	342	172.2	391	199.4	440	226.7
245	118.3	294	145.6	343	172.8	392	200.0	441	227.2
246	118.9	295	146.1	344	173.3	393	200.6	442	227.8
247	119.4	296	146.7	345	173.9	394	201.1	443	228.3
248	120.0	297	147.2	346	174.4	395	201.7	444	228.9
249	120.6	298	147.8	347	175.0	396	202.2	445	229.4
250	121.1	299	148.3	348	175.6	397	202.8	446	230.0
251	121.7	300	148.9	349	176.1	398	203.3	447	230.6
252	122.2	301	149.4	350	176.7	399	203.9	448	231.1
253	122.8	302	150.0	351	177.2	400	204.4	449	231.7

Incompatible Chemicals

The substances in the left-hand column must be stored and handled so that they cannot come into any contact with the substances in the right-hand column.

Alakaline and alkaline-earth metals, such as sodium, potassium, cesium, lithium, magnesium, calcium, aluminum	Carbon dioxide, carbon tetrachloride, and other chlorinated hydrocarbons. (Also prohibit water, foam, and dry chemical on fires involving these metals.)
Acetic acid	Chromic acid, nitric acid, hydroxyl-containing compounds, ethylene glycol, perchloric acid, peroxides, and permanganates.
Acetone	Concentrated nitric and sulfuric acid mixtures.
Acetylene	Chlorine, bromine, copper, silver, fluorine, and mercury.
Ammonia (anhydr)	Mercury, chlorine, calcium hypochlorite, iodine, bromine, and hydrogen fluoride.
Ammonium nitrate	Acids, metal powders, flammable liquids, chlorates, nitrites, sulfur, finely divided organics or combustibles.
Aniline	Nitric acid, hydrogen peroxide.
Bromine	Ammonia, acetylene, butadiene, butane and other petroleum gases, sodium carbide, turpentine, benzene, and finely divided metals.
Calcium carbide	Water (See also acetylene.)
Calcium oxide	Water.
Carbon, activated	Calcium hypochlorite.
Copper	Acetylene, hydrogen peroxide.
Chlorates	Ammonium salts, acids, metal powders, sulfur, finely divided organics or combustibles.
Chromic acid	Acetic acid, naphthalene, camphor, glycerol, turpentine, alcohol, and other flammable liquids.
Chlorine	Ammonia, acetylene, butadiene, butane and other petroleum gases, hydrogen, sodium carbide, turpentine, benzene, and finely divided metals.

Chlorine dioxide	Ammonia, methane, phosphine, and hydrogen sulfide.
Fluorine	Isolate from everything.
Hydrocyanic acid	Nitric acid, alkalis.
Hydrogen peroxide	Copper, chromium, iron, most metals or their salts, any flammable liquid, combustible aniline, nitromethane.
Hydrofluoric acid, anhydrous (hydrogen fluoride)	Aqueous or anhydrous ammonia
Hydrogen sulfide	Fuming nitric acid, oxidizing gases.
Hydrocarbons (benzene, butane, propane, gasoline, turpentine, etc.)	Fluorine, chlorine, bromine, chromic acid, sodium peroxide.
Iodine	Acetylene, anhydrous or aqueous ammonia.
Mercury	Acetylene, fulminic acid, ammonia.
Nitric Acid (conc)	Acetic acid, aniline, chromic acid, hydrocyanic acid, hydrogen sulfide, flammable liquids, flammable gases, and nitritable substances.
Nitroparaffins	Inorganic bases.
Oxygen	Oils, grease, hydrogen, flammable liquids, solids or gases.
Oxalic acid	Silver, mercury.
Perchloric acid	Acetic anhydride, bismuth and its alloys, alcohol, paper, wood, grease, oils.
Peroxides, organic	Organic or mineral acids; avoid friction.
Phosphorus (white)	Air, oxygen.
Potassium chlorate	Acids (See also chlorate.)
Potassium perchlorates	Acids (See also perchloric acid.)
Potassium permanganate	Glycerol, ethylene glycol, benzaldehyde, sulfuric acid.
Silver	Acetylene, oxalic acid, tartaric acid, fulminic acid, ammonium compounds.
Sodium	See alkaline metals.
Sodium nitrate	Ammonium nitrate and other ammonium salts.

Sodium oxide	Water.
Sodium peroxide	Any oxidizable substance, such as ethanol, methanol, glacial acetic acid, acetic anhydride, benzaldehyde, carbon disulfide, glycerol, ethylene glycol, ethyl acetate, methyl acetate, and furfural.
Sulfuric acid	Chlorates, perchlorates, permanganates.
Zirconium	Prohibit water, carbon tetrachloride, foam, and dry chemical on zirconium fires.

Safety in the Laboratory or Home Workshop

It is necessary to learn:
  Use of laboratory fume hoods
  Handling flammable solvents
  Mixing acids
  Glass blowing

  Common electrical hazards
  First aid (for four situations only)
      Stoppage of breathing
      Profuse bleeding
      Chemical burns (water only)
      Fire in clothing

  Use of portable fire extinguishers
  Compressed or flammable gases
  Handling and storing dangerous chemicals, including alkali metals.

An outstanding deficiency pertaining to laboratory safety seems to be a lack of awareness of hazards among nontechnical personnel. It is conceivable that increased emphasis on "briefing" custodial workers about the dangers of the laboratories in which they work, and periodic review of these conditions could substantially reduce the hazard of ignorance.

Third, a more universal use of safety glasses, reaction shields, and other personal protective devices seems to be needed. From the responses received, an increased program of education on the hazards of common laboratory procedures and the use of personal protective equipment to lessen these hazards would be helpful.

Chemical Hazards

All laboratories, whether they be biological, chemical, or radiological, utilize hazardous chemicals. The hazard may result from utilizing the "raw" product or from products of a chemical reaction between two or more substances or breakdown products developed through heating or aging. Laboratory personnel should have an acquaintance at least with the modes of entry, the physiological responses, both acute and chronic, and methods of roughly assessing the hazards potential of chemicals they are using.

Electrical Hazards and Management

The problem of handling electricity is probably one of the most ignored facets of safety, yet each year many needless deaths and injuries are caused through carelessness in handling even low voltages. It is also of importance to recognize that electrical equipment can act as an ignition source to activate a fire or explosion. Static electricity should be considered in this category.

Pressure Hazards

Pressure equipment, either high or low (vacuum), is a part of most laboratories. High-pressure apparatus such as gas cylinders, if improperly handled, can be very dangerous. This is especially true of oxygen. Precautions are necessary in handling, transporting, and in storing. Vacuum equipment, through implosions, can be every bit as dangerous as high-pressure explosions.

Cryogenic Hazards

Cryogenics or the use of low-temperature refrigerants requires a knowledge of the behavior of these materials under laboratory conditions. It is impossible to understand the design of a piece of cryogenic equipment or cryogenic experiment without an appreciation for the principles of insulation or the significance of extremely low temperatures. Misuse can result in severe injury.

Flammable Chemicals-Hazards

Fires and explosions account for the most dangerous and the most expensive types of laboratory accidents. A knowledge of the flammable properties of chemicals along with an understanding of potential sources of ignition is extremely vital. Storage and handling of these materials also requires special attention.

General Safety Considerations

A number of accidents and injuries in laboratories could very well result from improper lifting, falls, and lacerations from improper handling of glassware. Preventive measures in these areas are worthy of mention.

Ventilation

The principal method of hazards control in laboratory involves the effective use of ventilation, both general and exhaust. An example of exhaust ventilation is the fume hood which if improperly designed or used fails to give the desired protection. Observations indicate that the function of this equipment is not entirely understood and a number of misuses have been witnessed.

Laboratory Sanitation

Poor laboratory sanitation practices may be the cause of contaminating potable water supplies through temporary cross connections. At times, poor housekeeping practices may be dangerous because of blocking passages or by providing tripping hazards. Many chemicals are kept well beyond their usefulness, causing containers to deteriorate and leak, or chemicals to become unstable. Disposal of flammable and toxic chemicals also presents a problem.

Protective Equipment

All laboratories require that protective equipment of one type or another be immediately available. These devices may include eye wash, emergency shower, safety glasses, eye shields, protective clothing, and respiratory protection. Knowledge of the proper usage and limitations of such equipment is extremely important. At times, injury or death may result from improper selection and application of protective equipment.

Reports and Records

Reports and records are necessary adjuncts to any safety program and should be complete, accurate, and disseminated to the appropriate administrators. Accident reports are of little value unless periodically examined and tabulated in order to obtain a picture of local and overall problems.

Emergencies

The initial procedures one follows in an emergency oftentimes determine the ultimate outcome of the accident, both to the individuals and to the installation. The rudiments of first aid, fire fighting and reporting are vital. Personnel have to be continually instructed on procedures for medical and file emergencies and how and where to make these initial contacts. Such procedures are critical, especially when working alone.

Contact Lenses

It is important to wear eye protection in the chemical laboratory and expecially when wearing contact lenses.

Danger in Handling Acid

The heat evolution of the solution could have provided sufficient thermal shock to the glass to permit it to crack when lifted free of the counter, or setting it on a cold counter top; or a shock in setting it down, could have contributed to the bottom separating when the jug was lifted.

Written procedures for handling of acids should always be followed. Personal protective equipment consisting of face protection, rubber apron, and gloves are a necessity for this operation.

You Must Have Fire Extinguishers

If a fire breaks out in your office or apartment, get out fast. Many people are killed because they don't realize how fast a small fire can spread.

If you are caught in smoke take short breaths, breathe through your nose, and crawl to escape. The air is better near the floor.

Head for stairs – not elevator. A bad fire can cut off the power to elevators. Close all doors and windows behind you.

If you are trapped in a smoke-filled room, stay near the floor, where the air is better. If possible, sit by a window where you can call for help.

Feel every door with your hand. If it's hot don't open. If it's cool, make this test: open slowly and stay behind the door. If you feel heat or pressure coming through the open door, slam it shut.

If you can't get out, stay behind a closed door. Any door serves as a shield. Pick a room with a window. Open the window at the top and bottom. Heat and smoke will go out the top. You can breathe out the bottom.

DON'T fight a fire yourself.

DON'T jump. Many people have jumped and died without realizing rescue was just a few minutes away.

If there is a panic for the main exit, get away from the mob. Try to find another way out. Once you are safely out, *DON'T* go back in. Call the Fire Department immediately. Use alarm box or telephone. DIAL 911

If you find smoke in an open stairway or open hall, use another pre-planned way out.

REMEMBER: Get out fast. Don't underestimate how fast a small fire can spread. Use stairs, not the elevator. Close all doors behind you. Don't panic. Once you are safely out, call the Fire Department. Dial 911 or use alarm box. Don't go back in.

Trademark Chemical Manufacturers

The following is a list of Trademark Chemical Manufacturers. The names of these manufacturers are preceded by a number. In the list of Trademark Chemicals that follows this one, the manufacturers are referred to by a number alone.

1.	Abbott Laboratories, Inc.	North Chicago, Ill.
2.	Advance Solvents	New Brunswick, N.J.
2A.	Air Reduction Chem. and Carbide Co.	New York, N.Y.
3.	Air Products and Chemicals, Inc.	Phila., Pa.
4.	Alcan Sales Inc.	New York, N.Y.
5.	American Cyanamid Co.	Wayne, N.J.
6.	American Mineral Spirits	Palatine, Ill.
6A.	American Zinc Co.	Columbus, Ohio
6B.	Archer-Daniels-Midland Co.	Minneapolis, Minn.
7.	Arizona Chem. Co.	New York, N.Y.
8.	Ashland Chem. Co.	Columbus, Ohio
9.	Atlas Chem. Ind. Inc.	Wilmington, Del.
10.	Baker Castor Oil Co.	Bayonne, N.J.
11.	BASF Corp.	Paramus, N.J.
12.	Bordon Chemical Co.	Leominster, Mass.
13.	Bradco Co.	Los Angeles, Calif.
14.	Buckman Laboratories, Inc.	Memphis, Tenn.
15.	Burgess Pigment Co.	Sandersville, Ga.
16.	Cabot Corp.	Boston, Mass.

17. Calgon Corp. . . . . Pittsburgh, Pa.
17A. Carbide and Carbon Chemicals Co. . . . . New York, N.Y.
18. Carbola Chem. Co. . . . . Natural Bridge, N.Y.
19. Cargill, Inc. . . . . Phila., Pa.
20. Celanese Resins Corp. . . . . Louisville, Ky.
21. Central Soya Co. . . . . Chicago, Ill.
22. Chem. Manuf. Co., Inc. . . . . Stamford, Conn.
23. Ciba Prod. Co. . . . . Fair Lawn, N.J.
24. Colloids, Inc. . . . . Newark, N.J.
25. Columbian Carbon Co. . . . . New York, N.Y.
26. Commercial Solvents Corp. . . . . New York, N.Y.
28. Davison Chem. Co. . . . . Baltimore, Md.
29. Day-Glo Color Corp. . . . . Cleveland, O.
30. DeGussa, Inc. . . . . New York, N.Y.
31. DeLore Div. (National Lead Co.) . . . . New York, N.Y.
32. Dewey and Almy Chem. Div.(W.R.Grace & Co.). . Cambridge, Mass.
33. Dexter Chem. Corp. . . . . Bronx, N.Y.
34. Diamond Shamrock Chem. Co. . . . . Cleveland, Ohio
35. Dow Chem. Co. . . . . Midland, Mich.
36. E. I. duPont de Nemours and Co. . . . . Wilmington, Del.
37. Eastman Chem. Prod. Inc. . . . . Kingsport, Tenn.
38. English Mica Co. . . . . Stamford, Conn.
39. Farnow Inc. . . . . South Kearney, N.J.
40. FMC Corp. (Inorganic Chem. Div.) . . . . New York, N.Y.
41. Freeport Kaolin Co. . . . . New York, N.Y.
42. GAF Co. . . . . New York, N.Y.
43. General Electric Co. Schenectady, N.Y.
44. General Mills Chem. Div. . . . . Minneapolis, Minn.
45. Goodyear Tire and Rubber Co. . . . . Akron, Ohio
46. Glidden-Durkee Div. (SCM Corp.) . . . . Jacksonville, Fla.
47. B.F. Goodrich Chem. Co. . . . . Cleveland, O.
48. W.R. Grace and Co. . . . . New York, N.Y.
49. Great Lakes Carbon Corp. . . . . New York, N.Y.
50. Guardian Chem. Co. Hauppauge, N.Y.
51. Harshaw Chem. Co. (Div. of Kewanee Oil Co.) . . . . Cleveland, O.
52. Hercules, Inc. . . . . Wilmington, Del.
53. J.M. Huber Corp. . . . . Edison, N.J.
54. Humble Oil and Refining Co. . . . . Houston, Texas
55. Imperial Color Chem. . . . . Glenn Falls, N.Y.
56. Industrial Minerals . . . . Canada

57.	International Talc Co.	New York, N.Y.
58.	Jennat Corp.	Torrance, Calif.
59.	Johns-Manville	New York, N.Y.
59A.	Kessler Chemical Co.	Phila., Pa.
60.	Koppers Co.	Pittsburgh, Pa.
60A.	Krumbhaar Chemicals, Inc.	South Kearny, N.J.
61.	3M Co.	St. Paul, Minn.
62.	Malvern Minerals	Hot Spring National Park, Ark.
62A.	Marbon Chem. Div.	Washington, W. Virginia
63.	McCloskey Varnish Co.	Phila., Pa.
64.	McWhorter Chem. Co. Div.	Chicago, Ill.
65.	Merck and Co.	Rahway, N.J.
66.	Metalsalts Corp.	Hawthorne, N.J.
66A.	Metals Disintegrating Co.	Elizabeth, N.J.
67.	Michigan Chem. Co.	Chicago, Ill.
68.	Mineral Mining Corp.	Minneola, N.Y.
69.	Minerals and Chemicals Div.	Edison, N.J.
70.	Mobil Oil Corp.	New York, N.Y.
71.	Mobay Chem. Co.	Pittsburgh, Pa.
72.	Monsanto Corp.	St. Louis, Mo.
73.	Mooney Chem. Inc.	Cleveland, O.
74.	Naftone Inc.	New York, N.Y.
75.	National Lead Co.	New York, N.Y.
76.	National Starch and Chem. Corp.	Plainfield, N.J.
77.	Neville Chem. Co.	Anaheim, Calif.
78.	New Jersey Zinc Co.	Palmerton, Pa.
79.	NL Industries, Inc.	Heightstown, N.J.
80.	Nopco Chem. Div.	Newark, N.J.
81.	Nuodex Products Co.	Piscataway, N.J.
82.	Penn. Glass Sand Corp.	Pittsburgh, Pa.
82A.	Penn. Industrial Chem. Corp.	Clairton, Pa.
83.	Pennsalt Chem. Corp.	Phila., Pa.
84.	Pennwalt Corp.	Phila., Pa.
85.	Chas. Pfizer and Co.	New York, N.Y.
85A.	Pierce Oil Products	Rochester, N.Y.
86.	Proctor and Gamble	Cincinnati, O.
87.	Raybo Chem. Co.	Huntington, West Virginia
88.	Reichhold Chem. Inc.	White Plains, N.Y.
89.	Resyn Corp.	Linden, N.J.
90.	Rohm and Haas Co.	Phila., Pa.

91. Shanco Plastics and Chem. Inc. . . . . Tonawanda, N.Y.
92. Shell Chem. Co. . . . . New York, N.Y.
93. Shell Oil Co. . . . . New York, N.Y.
94. Sherwin Williams Co. Cleveland, O.
94A. Spencer Kellogg and Sons . . . . Buffalo, N.Y.
95. A.E. Staley Mfg. Co. . . . . Decatur, Ill.
95A. Standard Oil Co. of Cal. . . . . San Francisco, Cal.
96. Stepan Chem. Co. . . . . Northfield, Ill.
97. Thibaut and Walker Co. . . . . Newark, N.J.
97A. Thiokol Corp. . . . . Trenton, N.J.
98. Thompson, Weinman and Co. . . . . Cartlesville, Ga.
99. Titanium Pigment Corp. . . . . Heightstown, N.J.
100. Troy Chem. Corp. . . . . Newark, N.J.
101. Ultra Adhesives, Inc. . . . . Patterson, N.J.
101A. Union Bag and Paper Corp. . . . . New York, N.Y.
102. Union Camp Corp. New York, N.Y.
103. Union Carbide Corp. . . . . New York, N.Y.
104. United Sierra Div. (Cyprus Mines Corp.) . . . . . Los Angeles, Calif.
105. U.S. Industrial Chem. Co. (Div. of Nat'l Distillers and
      Chem. Corp.) . . . . New York, N.Y.
106. R.T. Vanderbilt Co., Inc. . . . . New York, N.Y.
106A. Velsicol . . . . Chicago, Ill.
107. Victor Chem. Co. . . . . Chicago, Ill.
108. Wallace and Tiernan Inc. . . . . Buffalo, N.Y.
109. Whittaker Clark and Daniels, Inc. . . . . New York, N.Y.
109A. C.K. Williams and Co. Easton, Pa.
110. Wyandotte Chemical Corp. . . . . Wyandotte, Mich.

Trademark Chemicals

INDEX

411

Heat-resistant metal topcoat, acrylic, 385

Heat-resistant primer, alkyd, 353–354

High gloss baking enamel, industrial, water-reducible, polyester, 389

High gloss, fast-drying enamel-interior, exterior, 62

High solids brushing masonry sealer, clear, epoxy-polyamide, 174

High solids epoxy enamel, beige, 218–219
    blue, 217–218

High-speed baking primer, alkyd, gray, 368–369

House paint, 24–25, 27, 30–33
    acrylic, 30–32
    gloss white, 25
    latex, 36
    ranch red, 32–33
    stain and blister-resistant, 27
    white, 24–25

Ignition waterproofing seal coating, acrylic, 331

Implement enamel, 109–111

Incompatible chemicals, 396–399

Ink, gravure, fluorescent, acrylic, 333

Industrial, acrylic baking enamel, white, 382–384

Industrial, baking enamel, clear, alkyd, 369–371
    alkyd-epoxy, 371–373

Industrial baking enamel, high gloss, water-reducible, polyester, 389

Industrial, epoxy, baking enamel, flat white, 377–379

Industrial, epoxy gloss enamel, amine, converted, 376

Industrial, epoxy-phenolic, chemical-resistant baking finish, clear, 375

Industrial, epoxy-polyurethane finish coat, 373–374

Industrial, epoxy, unpigmented baking finish, 374

Industrial, epoxy-urea, baking flexible finish, gray, 373

Industrial, epoxy-urea, brass, fast baking finish, 379

Industrial maintenance, primer, epoxy, stainless steel, 124

Industrial, metal coating, acrylic-vinyl, 382

Industrial paints, 350–389

Industrial, polyurethane, stain-resistant baked enamel, one-component, 380–381

Inhibiting pigment, for airless spray, 92–93

Interior, acrylic enamel, 275–279
    flat, 248–254

Interior, alkyd enamel, 263–268

Interior, alkyd, flat wall paint, 234–238

Interior butadiene-styrene, flat, 245–248

Interior, epoxy enamel, 268–272

Interior-exterior enamel, alkyd fast-drying, high-gloss, 62

Interior, fluorescent vinyl colors, 261

Interior-exterior, vinyl fluorescent flat, 262

Interior, latex flat wall paint, 239–242

Interior lumber primer, alkyd, fast-dry, 197

Interior mill white gloss enamel, triamino-ester, 142

Interior paints, 234–298

Interior, polyvinyl acetate flat, 243–244

Interior, polyvinyl acrylic latex enamel, 272–275

Interior, vinyl-acrylic emulsion flat, 254–260

International atomic weights, 391–393

Intumescent fire retardant coating, 328

www.ingramcontent.com/pod-product-compliance
Lightning Source LLC
Chambersburg PA
CBHW021025210326
41598CB00016B/911